Jackson Smith Schultz, Theron Skeel

The Leather Manufacture in the United States

A Dissertation on the Methods and Economies of Tanning

Jackson Smith Schultz, Theron Skeel

The Leather Manufacture in the United States
A Dissertation on the Methods and Economies of Tanning

ISBN/EAN: 9783337106478

Printed in Europe, USA, Canada, Australia, Japan

Cover: Foto ©berggeist007 / pixelio.de

THE
LEATHER MANUFACTURE
IN THE
UNITED STATES;

A DISSERTATION ON THE METHODS AND ECONOMIES OF TANNING.

BY JACKSON S. SCHULTZ.

WITH NUMEROUS ILLUSTRATIONS.

TO WHICH IS ADDED A REPORT ON THE RELATIVE ECONOMIES OF BURNING WET SPENT TAN, BY THERON SKEEL, C. E.

NEW YORK:
"SHOE AND LEATHER REPORTER" OFFICE,
1876.

PREFACE.

The proprietor of the SHOE AND LEATHER REPORTER proposes to meet a very general demand for information in regard to the manufacture of bark tanned leather by grouping together in permanent form a series of articles on the subject, written for the paper by Mr. JACKSON S. SCHULTZ. To make the work acceptable to the tanners, not only of America but of foreign countries, it will be appropriately illustrated, presenting in outline, and in some instances in working drawings, the principal mechanical inventions employed in the business, so that the machines may be duplicated and the methods of their use easily comprehended. It was found inconvenient to illustrate the work during its publication in serial numbers, but now that it is to appear in book form that feature will be supplied, and will add to its appearance and usefulness.

If the undertaking should be adequately encouraged by the tanners and finishers of bark leather, it will be followed at an early day by essays treating of "Sumac Tanning," and such other departments of leather and its fabrics as may be deemed worthy of consideration.

NEW YORK, *July*, 1876.

CONTENTS.

PAGE.

INTRODUCTION, - - - - - - - - - - xiii

CHAPTER I.

SELECTION AND CLASSIFICATION OF HIDES AND SKINS.

SIMILARITY OF STOCK AS TO WEIGHT, SUBSTANCE AND CONDITION, HIGHLY IMPORTANT—ASSORTING IN THE SOAK AND HIDE MILL—HIDES SHOULD BE OF EQUAL CONDITION ON ENTERING THE HANDLERS—BREAKING THE NERVE IN CALF, KIP AND UPPER LEATHER—IMPOLICY OF WORKING SEVERAL KINDS OF HIDES IN ONE YARD AT THE SAME TIME. - - - - - - - - - 17

CHAPTER II.

PREPARATION OF HIDES FOR THE BARK—SWEATING.

COLD AND WARM SWEATING—CONSTRUCTION OF SWEAT PITS—CLEANLINESS, LIGHT AND IMPENETRABILITY TO AIR NECESSARY IN COLD SWEAT PITS—PROPER CONDITION OF STOCK BEFORE ENTERING THE PITS—CARE NECESSARY IN THE SWEATING PROCESS—TREATMENT AFTER THE HIDES COME FROM THE PITS—PART LIMING AND PART SWEATING—GREASE AND SALT ON HIDES. - - - - - 23

CHAPTER III.

PREPARATION OF HIDES FOR THE BARK—LIMING.

GETTING READY THE LIMES—THEIR AGE AND STRENGTH—EFFECT OF THE LIME ON THE FIBER—PLUMPING AND BATING—PROF. LUFKIN'S PROCESS—THE "BUFFALO" METHOD—THEIR RESULTS—HANDLING IN THE LIMES. - - - - - - - - 31

CHAPTER IV.

PREPARATION OF HIDES FOR THE BARK—FLESHING AND TRIMMING.

THE BEAM WORK—CLOSE FLESHING, WITHOUT BREAKING THE GLUE CELLS—FLESHING LIME SLAUGHTER STOCK—WORK TO BE DONE BEFORE LIMING—FLESHING SWEAT STOCK—IT SHOULD BE DONE WITH A WORKER—THE GERMAN FLESHER—TRIMMING—CROP LEATHER AND BACKS—ADVANTAGES OF TRIMMING UPPER AS WELL AS SOLE—"ROUNDING"—TRIMMING BEFORE TANNING—BEST METHOD OF UTILIZING THE HEAD, PATE, ETC. - - - - 38

CHAPTER V.

GRINDING BARK—BARK MILLS.

THE INEXPENSIVE AND ABUNDANT POWER OF SOLE LEATHER TANNERIES—GRINDING BARK FINE AND UNIFORM—USEFULNESS OF SCREENING THE GROUND BARK—THE DOUBLE-GRINDING MILL—ITS EFFECTIVENESS WITH DAMP BARK—THE ALLENTOWN MILL—ADVANTAGE OF A STRONG MILL AND A WEAK COUPLER—THE SAW CUTTING MILL—A BARK CRUSHING MACHINE—THE PROPER SPEED AT WHICH MILLS SHOULD BE RUN—TANNIN LEFT IN THE BARK. - 46

CHAPTER VI.

LEACHING.

TANNIN VS. RESINOUS AND COLORING MATTER—TANNIN REQUIRED TO MAKE GOOD WEIGHT—EXTREME HEAT IN LEACHING INJURIOUS—FLOODING THE BARK—THE PRESS LEACH—HEAT TO BE APPLIED ONLY TO THE WEAKEST LEACH—CONSTRUCTION OF LEACHES—CLAY AND LOAM PACKING FOR THE SIDES AND BOTTOMS—WORKING THE PRESS LEACH—THE SPRINKLER LEACH—REVOLVING DETACHED LEACH. - - - - - - - - - 55

CHAPTER VII.

HANDLING.

THE HAND REEL—THE ROCKER HANDLER—ITS CONSTRUCTION AND OPERATION—IMMERSED DRUM WHEELS—A METHOD OF RAISING HIDES FROM THE HANDLER VATS—THE TUB WHEEL HANDLER—HANGING HIDES IN THE HANDLERS—THE "ENGLAND" WHEEL—HANDLING WITH THE COX ROLLERS. - - - - - 68

CHAPTER VIII.

HANDLING AND PLUMPING.

THE USE OF VEGETABLE AND MINERAL ACIDS—THE EARLY USE OF VITRIOL BY AMERICAN TANNERS—CONSIDERATIONS AFFECTING THE AMOUNT WHICH MAY BE USED—ITS EFFECT ON LIMED AND SWEAT STOCK—STRENGTH AND AGE OF LIQUORS TO BE USED IN THE HANDLERS—DIFFERENCES IN THE HANDLING AND PLUMPING OF SOLE AND UPPER LEATHER. - - - - - - - 76

CHAPTER IX.

LAYING AWAY.

TIME REQUIRED AND STRENGTH OF LIQUOR WHICH SHOULD BE EMPLOYED—TANNING IN THE HANDLERS VS. LAYING AWAY—EUROPEAN METHODS—"BLACK ROT" AND WHITE SPOTS—THEIR CAUSES AND THE REMEDIES—SHOULD HIDES BE LAID AWAY GRAIN UP OR FLESH UP?—MAKING WEIGHT IN THE LAST LAYER. - - 84

CHAPTER X.

DRYING AND FINISHING.

WASHING AND SCRUBBING THE LEATHER—THE "HOWARD SCRUBBER" —WHEEL OR DRUM SCRUBBING—DRAINING—HOW THE ADMISSION OF LIGHT AND AIR SHOULD BE REGULATED IN DRYING—DAMPENING BEFORE ROLLING—THE FIRST AND SECOND ROLLING—EFFECT OF THE ROLLING ON THE BUFFING QUALITIES—BLEACHING WITH SUGAR OF LEAD AND SULPHURIC ACID—THE WARM SUMAC BATH —EFFECT OF THE LATTER ON CALFSKINS, GRAIN LEATHER, ETC. - 94

CHAPTER XI.

THE CAUSES WHICH AFFECT COLOR AND ASSIST IN THE MAKING OF A VALUABLE EMBOSSING GRAIN.

WHY LEATHER SHOULD BE THOROUGHLY DRIED—STRUCTURE OF THE GRAIN—IMPORTANCE OF A PERFECT FINISH—CARE TO BE TAKEN TO AVOID STAINS AND DISCOLORATION—"CUIR" COLOR—THE NATURAL HEMLOCK COLOR—"RUSSIA LEATHER" COLOR—FRAUDS IN SELLING HEMLOCK FOR OAK LEATHER DURING THE WAR—COLORING TO BE DONE IN THE HANDLERS—EFFECT OF "STRIKING" THE GRAIN. - - - - - - 103

CHAPTER XII.

CONSTRUCTION OF TANNERIES—THE TURRET DRYER.

HOW THE ADMISSION OF LIGHT AND AIR IS CONTROLLED IN THE TURRET DRYER—ITS CAPABILITIES FOR DRYING LEATHER IN QUICKER TIME, WITHOUT REGARD TO THE WEATHER—ITS CONSTRUCTION, AND HOW ITS CAPACITY SHOULD BE PROPORTIONED TO THAT OF THE YARD—HOW AND WHEN HEAT SHOULD BE USED—HOW TO PREVENT DISCOLORATION OF THE LEATHER—SAVING OF LABOR IN THE TURRET DRYER. - - - - - - 110

CHAPTER XIII.

CONSTRUCTION OF TANNERIES—PLANS, FOUNDATIONS, ETC.

THOROUGH EXAMINATION OF PRESENT STRUCTURES AND APPLIANCES ADVISABLE BEFORE BUILDING—IMPORTANT CHANGES FROM THE PRESENT GENERAL USE OF STEAM INSTEAD OF WATER POWER—LOCATING ON "MANUFACTURING" AND "CULINARY" STREAMS—A LOAM, CLAY, OR SANDY FOUNDATION—FILLING IN BETWEEN VATS AND LEACHES WITH LOAM OR CLAY—PLACING THE VATS—THE "BUFFALO" VAT—"BOX" VATS—THE PROCESS OF "PUDDLING" IN SETTING THE VATS—UPPER CONDUCTORS—SIDE AND END WALLS. - - - - - - 117

CHAPTER XIV.

CONSTRUCTION OF TANNERIES—LEACHES.

ROUND OR SQUARE LEACHES—THE DURATION OF LEACHES ABOVE AND SUNK IN THE GROUND—HOW TO BUILD A ROUND LEACH—HOW TO MAKE AND SET LEACHES IN THE GROUND—THE CAPACITY OF THE SETS OF LEACHES TO BE PROPORTIONED TO THE SIZE OF THE TANNERY. - - - - - - - 126

CHAPTER XV.

CONSTRUCTION OF TANNERIES—FRAME WORK AND LOCATION OF BUILDINGS.

WHY THEY SHOULD BE ONLY ONE-STORY HIGH FOR THE YARD AND BEAM HOUSE—SAVING IN INSURANCE BY SEPARATING THE BUILDINGS—CONVEYING LEATHER TO THE "TURRET" DRYER—TRANSMITTING

POWER TO DISTANT BUILDINGS—PROPER SPEED FOR BARK MILLS AND ELEVATORS—SIMPLE PROVISIONS AGAINST FIRE AND BREAKAGE, AND TO PREVENT DUST. - - - - 134

CHAPTER XVI.

THE ROSSING OF BARK.

THEORIES OF THOSE WHO ADVOCATE ROSSING—ITS COST—DIFFICULTY OF ROSSING WITHOUT TOO GREAT LOSS OF TANNIN—STRENGTH OF LIQUORS WHICH MAY BE OBTAINED FROM ROSSED AND UNROSSED BARK—POSSIBLE ADVANTAGE IN ROSSING BARK FOR EXPORT IN THE "LEAF." - - - - - - 141

CHAPTER XVII.

UTILIZATION OF TANNERY REFUSE.

BURNING THE WET TAN—GLUE STOCK—IMPORTANCE OF KEEPING THE PIECES PURE AND SWEET—PRESERVING, CLEANSING AND DRYING THEM—USES FOR CATTLE HAIR—THAT WHICH COMES FROM SWEAT OR LIMED STOCK—WASHING, DRYING AND PACKING—FERTILIZING LIQUIDS FROM THE LIMES AND SOAKS. - - - 146

CHAPTER XVIII.

TANNING MATERIALS.

DIFFERENT KINDS OF HEMLOCK BARK—INFLUENCE OF SOIL AND CLIMATE ON THE QUALITY—HEAVY AND LIGHT BARK—VARIETIES OF OAK BARK—THE "SECOND GROWTH" BETTER THAN THE FIRST—GAMBIER—ITS GROWTH AND PREPARATION FOR MARKET—ITS COST COMPARED WITH THAT OF BARK—VALONIA, DIVI DIVI, MYRABOLAMS—"SWEET FERN," ETC. - - - - 155

CHAPTER XIX.

THE COST OF TANNING.

THE SEVERAL ITEMS VARYING WITH DIFFERENT TANNERS—DIFFERENCES FROM UNEQUAL WEIGHT OF THE CORD OF BARK—THE AMOUNT OF TANNIN IN UPPER LEATHER AS COMPARED WITH THAT IN SOLE LEATHER—COMPARATIVE COST IN MAKING HEAVY AND LIGHT GAINS—THE THEORETICAL STRENGTH OF BARK NEVER REALIZED—COST OF "UNION" AND OAK TANNING—ESTIMATED COST OF TANNING IN EUROPE. - - - - - 169

CHAPTER XX.

QUICK TANNING PROCESSES.

COMMON ERRORS OF THOSE OUTSIDE OF THE TRADE—HOW WORTHLESS PATENTS ARE MULTIPLIED—EXPERIMENT IN TANNING BY HYDROSTATIC PRESSURE—VACUUM TANNING—DIFFICULTIES ATTENDING THIS METHOD—HOW AGITATION OF THE FIBER FACILITATES THE PROCESS—A GENTLE MOVEMENT, WITH OCCASIONAL REST, MOST EFFICACIOUS—TANNING VS. TAWING. - - - 176

CHAPTER XXI.

THE SPECIES AND GROWTH OF HIDES.

"HEALTHY" AND "WELL GROWN" HIDES—DIFFERENCES IN HIDES AT VARIOUS SEASONS OF THE YEAR—EFFECT OF CLIMATE AND FOOD ON TEXTURE AND GROWTH—IMPROVED BREEDS OF CATTLE MAKE HIDES THIN AND SPREADY—COLD CLIMATE MAKES A COARSE FIBER AND WARM CLIMATE A FINE TEXTURE—EAST INDIAN, AFRICAN AND SOUTH AMERICAN HIDES—THE HIDES FROM THE EASTERN AND MIDDLE STATES AS COMPARED WITH THOSE FROM THE WESTERN PRAIRIES—CARE TAKEN OF CATTLE IN EUROPE. - - 188

CHAPTER XXII.

FRENCH AND GERMAN CALF AND KIP.

WHERE OUR IMPORTED STOCK COMES FROM—CAREFUL ASSORTING OF THE RAW STOCK TO INSURE UNIFORMITY IN WEIGHT, SUBSTANCE AND GENERAL CONDITION—SOAKING AND MILLING—BREAKING THE NERVE—LIMING—BATING AND WORKING OUT LIME—COLORING AND HANDLING—LAYING AWAY AFTER WORKING—STUFFING—DRYING—SLICKER WHITENING—BLACKING AFTER THE STOCK IS CUT OUT USUAL IN EUROPE—VEGETABLE OILS USED INSTEAD OF FISH OILS—DEFECTS IN FOREIGN CALFSKINS—STEADY IMPROVEMENT IN AMERICAN CALFSKINS. - - - 194

CHAPTER XXIII.

GRAIN AND BUFF LEATHER.

SPLITTING MACHINES—MAKING SPLIT LEATHERS FROM GREEN HIDES OR FROM TANNED LEATHER—EVERY KIND OF NATURAL GRAIN SUCCESSFULLY IMITATED—STRENGTH AND DURABILITY OF SPLIT

LEATHERS—THEIR INTRODUCTION TO EUROPEAN CONSUMERS—ESSENTIALS TO BE CONSIDERED IN THE MANUFACTURE OF GRAIN AND BUFF LEATHER. - - - - - - 210

CHAPTER XXIV.

CURRYING AND FINISHING.

THE STUFFING WHEEL AND HOW TO USE IT—TO PURIFY AND CLEANSE DIRTY GREASE—HOW TO MAKE STUFFING—FLESH BLACKING—FLOUR AND SIZE PASTES—HARM THAT MAY BE DONE BY DEPENDENCE UPON RECIPES—DAMPENING LEATHER BEFORE AND AFTER APPLYING OIL AND TALLOW. - - - - - - 218

CHAPTER XXV.

DIRECTIONS FOR THE CONSTRUCTION OF DETACHED FURNACES FOR BURNING WET SPENT TAN, BY THERON SKEEL, C. E. - - - 224

LIST OF ILLUSTRATIONS.

PAGE.

Cold Sweat Pits, - - - - - - - - - - - 249
Hide Mill, - - - - - - - - - - - 251
Hand Reel, - - - - - - - - - - - 253
Rocker Handler, - - - - - - - - - - 255
Sole Leather Roller, - - - - - - - - - 257
Turret Dryer, - - - - - - - - - - - 259
Sprinkler Leach, - - - - - - - - - - 261
Bark Cutting Mill, - - - - - - - - - 263
Allentown Bark Mill, - - - - - - - - 265
Howard Leather Washer, - - - - - - - 265
Salem Tan Press, - - - - - - - - - - 267
Lockwood Leather Scourer, - - - - - - - 269
Fitzhenry Leather Scourer, - - - - - - - 271
Burdon Leather Scourer, - - - - - - - 273
Stuffing Wheel, - - - - - - - - - - 275
Charles Korn's Whitener, - - - - - - - 277
Union Whitener and Buffing Machine, - - - - 279
Fisk's Whitener and Buffing Machine, - - - - - 281
Union Leather Splitter, - - - - - - - - 283
Henry Lampert's Hide Worker, - - - - - - 283
Outline and Trim of Hide, - - - - - - - 285
Tanners' and Curriers' Tools, - - - - - - - 287
Hoyt Furnace at Wilcox Penn., Plates I, II, III, - 289, 291, 293
Design for Wet Tan Furnace, Plates IV, V, VI, - 295, 297, 299

INTRODUCTION.

This Centennial year seems an appropriate period in which to review the progress and present attainment of the tanner's art in America. In the chapters which are to follow it is not proposed to give a chronological history so much as the general progress of the trade, and even that progress will be considered in subordination to the permanent advancement of the whole manufacture, rather than to the glorification of any special period of our trade history, or the commendation of any particular man or class of men, however conspicuous they may have been during critical periods. This purpose will make it unnecessary to inquire whether "Simon" was a good or poor tanner; indeed, whether he was a tanner at all, since, whatever his merits as an artisan, they have long since been surpassed by others with far better methods.

For similar reasons no attempt will be made to bring under review the various exploded patents and so-called improvements that have for the past fifty years been pressed on the attention of tanners. No volume, however large, would be adequate to contain the recital of their origin and pretensions. But so many of such improvements as have been at any time adopted by any considerable number of the trade—and particularly if they have a practical existence in any of our modern yards—will receive respectful consideration. This, to be useful, must be candid, and free from all bias.

To intelligently contrast the merits of mechanical inventions—to hold an even hand in weighing the advantages and disadvantages of systems and methods which include among their advocates men equally intelligent—is a task not likely to be accomplished without involving the writer in much censorious criticism. But if an entire absence of all ownership or interest in patented or other improvements can qualify one for the task, then these chapters will contain a fair presentation of the merits of all claims.

The form chosen in which to present this subject is intended to elicit honest inquiry and full discussion. Instead of a treatise giving in dogmatic language the processes which it is claimed tanners have adopted or must adopt, it is intended here to present the merits and defects of known systems and methods as their advocates would state them, and hold these up in contrast with other methods commended by equally intelligent men, and thus, by candid comparison, show their relative merits, and indicate such preference as may appear to be justified by our experience. This course of inquiry and presentation will, it is believed, lead to far better results than if preconceived and fixed theories and opinions were enunciated and defended. In still other words, these chapters will contain a tolerant discussion of the present condition of the tanner's art in America, and, by way of contrast and honorable emulation, with the relative position of the same trade in the more advanced nations of Europe, and in the treatment of these subjects there will be no pedantic use of terms, either mechanical or otherwise; only such phrases and words will be employed as are familiar to all practical tanners.

While it is conceded that the tanner's art is largely chemical in its nature, and may in the future be greatly advanced by the study and application of chemical laws, in this preliminary inquiry it is thought expedient for both writer and reader to confine attention to the tanner's art as understood by ordinary workmen.

It is proposed here to treat only of the subject of "tanning" as contradistinguished from "tawing." The art of "tanning" differs so widely from that of "tawing" that an expert in the one may be a novice in the other. They do not assimilate any more than the glove maker does with the boot or shoe maker. The latter uses "tanned," while the former uses "tawed" leather. Sir Humphrey Davy has drawn the distinction by the adoption of a chemical formula and test. He says, in substance, that leather is a chemical combination of gelatine and tannin, its characteristic being that, when combined, water will not separate the constituents or dissolve the connection; whereas, in the tawing process, water will separate them and return the gelatine and salt or alum into their original elements.

In America we have but little experience with any other than bark tannage, and, therefore, if we speak at all of the various substitutes for bark used by tanners in the old world, it must be from a very limited experience and a qualified knowledge. In deference to the success of the tanners of Great Britain we are compelled to admit that vegetable substances other than bark—such as cutch, terra japonica, valonia, myrabolams, divi divi, etc.—do tan leather. These materials practically do the tanning for Great Britain, and make a serviceable and even artistic leather. The barks of our forests, particularly those of the hemlock and the oak, are, then, only two among the tanning agents which are to be employed in making leather. But as these are the agents with which we are most familiar—indeed, the only agents in general use in this country—it is with reference to their use that these chapters will treat, considering only incidentally such other tanning materials as come into competition.

The only exception to the use of barks or bark extracts in this country is that of "terra japonica," and this is used only to a limited extent. But as this tanning substance has been the innocent cause of much misapprehension, and has led many novices into grave mistakes, as we pass along we may turn aside occasionally to point out this rock on which many hopes and even fortunes have been wrecked.

In illustrating the machinery which is most approved by tanners and workers in leather, it must be understood that only such kinds are here presented as are new, or comparatively so, in construction. Many useful machines are now in use, but so old and familiar that it is not thought expedient to encumber these pages with their presentation. Some of the illustrations are of old machines with new attachments; whenever this occurs an attempt will be made to designate the added novelty, in the text accompanying the drawing, or in the drawing itself.

CHAPTER I.

SELECTION AND CLASSIFICATION OF HIDES AND SKINS.

SIMILARITY OF STOCK AS TO WEIGHT, SUBSTANCE AND CONDITION, HIGHLY IMPORTANT—ASSORTING IN THE SOAK AND HIDE MILL—HIDES SHOULD BE OF EQUAL CONDITION ON ENTERING THE HANDLERS—BREAKING THE NERVE IN CALF, KIP AND UPPER LEATHER—IMPOLICY OF WORKING SEVERAL KINDS OF HIDES IN ONE YARD AT THE SAME TIME.

To secure the best results throughout the tanning and finishing of leather, there must be equality of condition, and the conditions must be favorable; among these conditions are similarity or equality of stock at the start. Dry flint hides cannot be worked with salted; heavy hides cannot be brought into condition with light ones in the same pack. As far as possible, then, both as to weight, substance and condition, the pack must be in all its respects equal. Some tanners think they effect the same result by assorting their packs after soaking, or after milling or sweating; but, exercising all the judgment that is possible from the outset, there will be opportunity to reject and assort in each stage of the process.

Where it is possible the whole lot of hides should be classified before any portion is put into the soaks. This is not always convenient or possible—as, for instance, where the hides are brought from a distant depot, sometimes many miles removed, on the return of teams from delivering loads of leather sent away. But where it is at all convenient the heavy hides should be selected and worked in first, enabling the whole lot to come out at the same time. This practice will be found far better than to assort the packs on the last layer, throwing the heavy sides back, and, aside from the consideration of being able to return the whole lot promptly and together, it enables the tanner to give the hides better and more considerate treatment in the soaks, mills and sweats.

After classifying all hides of the same weight and general appearance it will be found that some soften much more readily than others. This difference will be discovered generally in the first milling. The attendant, standing by the side of the mill constantly, with his hand on each side as it comes round in its turn, will "draw" (remove) the soft sides. This will leave the hard and unyielding sides in the mill until all the soft ones have been taken out. As a general experience it will be found inexpedient to "force" these remaining sides. It is better that they be taken out and returned to the soak, and not remilled until a future period, depending on the weather and condition of the water. When the weather is very cold these hard sides may be treated to a bath of moderately warm water—say up to 80° of heat—for a few hours. If they are sound they will usually yield on the second milling.

The packs should not be formed to go into the sweats until after they have come from the mill. The experience gained in milling will enable every intelligent man to send into the sweats a given number of sides in nearly equal condition. From the entrance of the stock into the reception, or hide house, until it gets into the sweats, and, indeed, after it has come from the sweats, the object should be to equalize the sides in the same pack.

Especial attention to properly softening is confined almost

exclusively to the dry flint stock; both green and dry salted hides soften without effort, and yet it is important that the substance and condition should be the same in each pack, even in this description of hide. Pickled hides should be kept separate from salted, and green salted from freshly taken off hides.

If it is important in sole leather hides to maintain uniformity of condition in the beam house, it is much more so in upper leather hides, kip and calfskins. The writer does not hesitate to say that, in the absence of this equality of condition, upper, harness and calfskins cannot be carried through the tanning and finishing processes in a workmanlike manner and with good results.

This difficulty may be aptly illustrated by the experience of many small tanners, who cannot wait until they "take in enough stock" of any one description to make a pack; they feel obliged to make up a pack of hides, kip and calf, some green, some fully dried, and some partially dried or dry salted. The result is that neither class is well worked or prepared; as they had the misfortune of being joined in the beam house, they go linked through the yard, and the whole pack comes out a tanner's abortion.

But even in cases where hides, kips and calfskins are worked separately in the same yard, the tanner often thinks it quite sufficient if he works each class by itself, whereas there are as many conditions to be observed in each class of salted as in dry hides; not, perhaps, in order to secure a sound result, but much more is required of this class of stock. The grain must be fully preserved, and the whole fiber must be reduced to a pulp—as it can be, when the nerve is thoroughly broken, and not before. This nerve depends upon many conditions for its tenacity. It is easiest broken when fresh from the animal, but may be severed under any ordinary circumstances. *No calf, kip or upper leather can be made with those yielding qualities so highly appreciated in our country without the severance of this nerve,** and the sooner our tanners appreciate this fact the sooner will they make leather to take the place of the French and German calfskins that so largely supply our best custom boot and shoe makers at the present time.

In all the processes, commencing with the soaking and milling, or wheeling, through the lime and bate, each pelt must be individually treated, and if the conditions are much varied, more judgment and care will be necessary in their treatment as a whole than if they are substantially alike.

In the latter case ordinary intelligence would suffice to perform creditable work; this degree of intelligence is all that the employer has a right to expect, and hence the importance of making as light drafts upon the brain-power of his men as possible, by making the labor uniform on each piece of stock.

How few calfskin tanners in this country think it important to classify their skins! Do they not work all skins from six pounds to twelve in the same pack? Whatever is classed as "veal" goes together; the first selection that is thought of is when the finishers are wanting stock; the packs are then

* So little is known in this country of the process of breaking the nerve, or even of the presence of such nerve, that I venture to add this note, somewhat out of the order in which this subject is treated. When a calf is first killed, this nerve will be seen, by close observation, twitching and contracting on the flesh for a few moments after death; the whole flesh appears alive with muscular action; when closely observed, this action will appear to be the result of innumerable interlacing nerves, which a wise Providence has placed there to expand or contract to meet the requirements of the seasons and the varying condition of the animal. When these nerves become quiet and fixed they hold the fiber of the skin, giving it a compact and even rigid feeling. To demonstrate the existence of this nerve, let the following experiment be tried: Take a green skin, throw it over the tanner's beam, and, with a worker, put ten minutes' hard work on the flesh side of one-half of the skin; the result will be that the half of the skin worked will be distended and soft—even pulpy. Much more would this be the effect if done after soaking thoroughly, as it should be before the usual beam work is done.

assorted and the tanned skins are selected out and sent to the currying shop, while the heavy ones are given another liquor. This is beginning at the wrong end; the selection should have taken place before the skins entered the beam house, when the advantages of classification would have been secured all the way through the process. In a well regulated calf or kip skin yard, from the time the skins enter the tannery they are mated (for reasons hereafter to be stated), and continue this connection through the whole after tanning process. But how can dissimilar sizes and substances be suitably paired, and so placed, grain to grain, as to fully cover each other?

What has been said thus far goes to the advantage of the intrinsic quality of the stock; but suppose some hides or skins are damaged, or partially so? These should by no means be allowed to contaminate the good. They are the sick members, and must be placed in hospital, under observation; they may not all have the same disease, and must be placed in different "wards" or "apartments" for special treatment.

When one thinks of the indiscriminate and forcing processes which valuable stock receives at the hands of many tanners, the inhumanity of the treatment is forced on one's mind. Sick or well, strong or weak, large or small, the same methods, the same trying ordeal, must be passed by all, and that so few should break and fail is the wonder.

It remains only for me to say a word about the impolicy of working a variety of hides in the same yard. It is not to be denied that some tanners succeed in making good stock out of a variety of hides under treatment at the same time; but this is the exception, and should not be ventured upon by the average tanner. At least one season's or one year's hides should be of one kind, or as nearly so as possible. Buenos Ayres, Montevideo and Rio Grande are sufficiently alike to be classed together; Central American and Matamoras, and even dry Texas, are, possibly, similarly conditioned. California and Western may well be treated as similar hides, requiring like treatment; but there cannot safely be treated dry salted and dry flint hides in the same beam house; lime and sweat stock cannot go through together without danger, or certainly with hope of the most satisfactory results. The best leather is made by tanners who work a uniform description of hide. This is the usual experience, and is based on common sense.

CHAPTER II.

PREPARATION OF HIDES FOR THE BARK—SWEATING.

COLD AND WARM SWEATING—CONSTRUCTION OF SWEAT PITS—CLEANLINESS, LIGHT AND IMPENETRABILITY TO AIR NECESSARY IN COLD SWEAT PITS—PROPER CONDITION OF STOCK BEFORE ENTERING THE PITS—CARE NECESSARY IN THE SWEATING PROCESS—TREATMENT AFTER THE HIDES COME FROM THE PITS—PART LIMING AND PART SWEATING—GREASE AND SALT ON HIDES.

In contradistinction to the sweating process of Southern Europe and Great Britain, the American method is called the "cold sweat"; theirs is denominated the "warm sweat." In France, Germany, Austria and Switzerland, perhaps also in other countries of Continental Europe, the tanners sweat their green hides by piling one on top of the other, laid out flat, and covering them up with spent tan or horse manure until decomposition begins. This is their process of preparing sole leather hides; for upper they lime, substantially as we lime our upper stock.

In this chapter both the construction and practical operation of the sweat pits will be considered. The modifications and changes of construction in the form of sweat pits for tanners have kept pace with the alterations which have been made in the building and location of ice houses for the people at large. The same principles govern both. A good sweat pit would make a good ice house; possibly a good ice house might not make a good sweat pit—but only for the reason that it would not be sufficiently commodious and controllable.

In Great Britain it is customary to "steam sweat" sheepskins until the wool yields. This is done by inserting steam under a false bottom of a chamber, hung up in which are the skins with wool on. The two methods mentioned may be regarded as "warm sweating" in contrast with our system of "cold sweating." It may then with truth be claimed that our method of sweating is peculiar to this country.

It is now fully demonstrated that a wood, brick or stone structure on the top of the ground can be so completely protected from the rays of the sun and other atmospheric influences as to make a good sweat pit. The ice companies have adopted surface structures of wood, "filled in" with saw dust, tan bark or charcoal between the outside clapboards and the inside lining of their buildings, and this same form of structure will make a most serviceable tanners' sweat pit. But since the sweat pit is subject to greater changes of atmosphere than the ice house, it is desirable that the inner lining of the sweat pit should be of a more enduring substance than wood. The damp but warm atmosphere of a tanner's sweat pit decomposes the fiber of the wood very fast—a sound hemlock plank thus exposed becoming worthless in a few years. On account of this liability to decay, if for no other reason, the sweat pits of tanners should be constructed of stone or brick. But these structures may be wholly above ground, and should be so placed that a wheelbarrow may be run from the floor of the beam house into the main passage way of the sweat pit. These passage ways should be wide—not less than six to eight feet—and so thoroughly lighted, both from top and ends, as to make the passage through them by the workmen both easy and agreeable. The hight of the main passage way should extend above the surrounding pits, and by a "lantern" construction of the roof both light and air can be secured in the passage ways below.

The pits themselves should extend from both sides of the main passage, and be connected with folding doors, wide enough, when fully open, to admit a wheelbarrow. Each of these pits should be large enough to contain one pack of hides, and high enough to admit the hanging of a side doubled, with a space of fully two feet above and one foot below the racks, on which or from which the sides are suspended by tenter hooks. The width of the pits should be about eight feet. This will give space for two tiers of sides and a small passage to enable the attendant to make his examinations. Both this passage and the room above and below the hanging sides give opportunity to introduce light and ventilation when required.

There is no reason why these pits should not be so light that close observations may be made without the use of a lamp, and they should always be kept in such cleanly and orderly condition as will permit the foreman or employer to enter and make inspections without fear of soiling their clothes. Too much stress cannot be laid on this feature of our more modern sweat pits. The fact that such inspections are liable to be made at any time keeps a most healthy check on the attendant.

The temperature of the sweat pits should be held under control by steam and cold water, with which the main passage should be amply supplied by means of pipes. A properly constructed pit should have a false bottom, under which the steam may be forced, to find its way in condensed spray up through the suspended sides. This process will adequately warm the pit. When too warm, cold water may be thrown from the mouth of a sprinkler over the whole surface, and thus, in a few moments, cool the whole space, and leave a desirable moist atmosphere. When in proper condition the pits will stand at a temperature of 60° to 70°, F., with globules of water collected on all parts of the suspended sides, occasionally dropping in their condensation from the ends of the hair.

Each one of the pits should have at least four lights, 7 by 9 inches, in the end, with a lantern ventilator in the top of the latter, which can be opened and closed by a cord. The pits should be covered on top by timbers flattened on three sides, and heavy enough to hold up at least two feet of earth. This earth, sodded over and kept well watered, will amply protect the pits from the rays of the sun, and the ammonia which arises from the pits will produce a most luxurious growth of grass, vines or flowers, on this earth covering. The sides of the pits should be protected by a banking of earth or spent tan. The former is the most desirable, for this earth may be made the means of cultivating grape or other vines with the most artistic and even serviceable effect.

With pits thus constructed and under proper control, hides may be properly sweat in from three to seven days—usually in about four or five days. To accomplish this result the hides or sides must be in proper condition.

For the most part only dry flint hides are prepared by the sweating process, in this country, and yet, as we proceed in our inquiry, possibly we may find that sweating in connection with liming may serve a most useful purpose, particularly for green or dry salted foreign hides. The native green salted hides are mainly taken by lime tanners, who find no occasion to use the sweating process.

Flint dry hides, or indeed any hides that enter the sweat pit, must first be made absolutely soft in all their parts—pates as well as skirts and butts. In this respect tanners do not exercise sufficient care. Where the hides are tender (liable to break) there is a strong temptation to stop short of this desirable condition. Probably less breaking thus results, but then "*old* grain" and "hard spots" will come in to offset the breaking. If at any time half prepared, hard sides, should go into the sweats, they should go in with companions equally delicate and tender in their constitution, and all should then be treated alike.

The sides should be hung on the racks by tenter hooks, either suspended from the shoulder or from the pate and butt; whichever way is adopted, the practice should be uniform, so that a uniform result may be anticipated.

As the sweating process advances with greater rapidity in the top than at the bottom of the pit, and as the thicker portions of the hide resist the action of the sweat longer than the thinner portions, it is desirable, as far as practicable, to hang the pates and butts higher than the shoulders and bellies, but as this is difficult to manage, the result is accomplished by changing the positions of the sides or hides as they advance in the process. Usually, after three or four days, the "assorting out" and changing of position commences, and on the fidelity with which this is done will damage be prevented and a good result secured.

No hides, however uniform in character, will sweat exactly alike, and it is for this reason that the careful attention of the employee must be secured. No hour in the whole day should pass without a visit to the advanced sweats. When a few sides give indications of "coming" prematurely, before their proper time, they should be dropped to the bottom of the pit, and allowed to lie in piles until their less advanced companions catch up in the process of decomposition.

Supposing all the sides to be in the same state of forwardness, the present practice is to throw them in the mill, and for a few moments wash out or off the slime, and rub the hair off, or so much of it as can be removed easily. During this short and damaging process, two things happen: 1. The loose hair is fulled into the flesh so firmly as to make it difficult to remove afterward on the beam, and, 2. Much of the gelatine of the hide is lost, as at this period it is in very nearly a soluble condition and will part from its proper lodgment in the fiber almost as freely as the slime and dirt with which the surfaces are covered; indeed, much of the substance that is regarded as "slime" and "dirt" is the gelatine, which, when combined with tannin, goes to make leather. There is no doubt that many of even our best tanners make too much dependence on this after milling to soften their stock, and with this end in view allow the mills to run on this tender stock far too long.

Some of our most thoughtful tanners have of late, to meet this difficulty, thrown the sides when they first come from the pits into a weak lime water, and allowed a slight reaction to take place. This action of the lime is indicated by a slight plumping of the side, and the disappearance of that slimy, slippery feel, which always attends sweat stock. Besides other savings, this after-liming prevents, to a great extent, the hair from attaching itself to the flesh while in the hide mill. But beyond these savings I venture to suggest that even the slight liming here indulged in will counteract the action of the vitriol in the handlers, and to those tanners who use vitriol this liming will be of service, for lime and vitriol will counteract and destroy each other. It is well known that on purely lime stock vitriol may be used moderately with advantage, both as a "bate" and as a means of plumping the leather.

It has been demonstrated that part liming and part sweating the same hides will answer a very excellent purpose. The writer has in his mind an experiment tried on a large scale, viz., on 10,000 Texas and New Orleans green salted hides, which produced 73 pounds of leather to the 100 pounds of hide, and the leather thus made was excellent in quality—both plump in offals and fine in fiber. If the present vitriol-raising process is to be continued, and finds favor, in my judgment this partial liming must be resorted to if a reasonably good buff is secured.

When we are assured, as a justification of our practice, that the tanners of Great Britain plump all their bends and butts with vitriol, we should remember that this description of leather is there made for the most part out of green salted hides from South America or Spain, which are *highly limed.* The English tanners occupy from eight to fourteen days in their liming process. The condition of the hide produced by this excessive liming will justify the use of vitriol without endangering the buff or fiber. With our sweat stock it is far different; our sweat sides contain nothing to neutralize this mineral acid, and the result is most disastrous to the buff, and also to the intrinsic vitality of the fiber.

There can be no doubt that our American system of cold sweating is calculated, beyond any other known method, to make a firm, compact fiber, when properly used, and, besides, it is especially adapted to the preparation of the dry hides of our continent.

It only remains for me to say a word on the subject of "grease" and "salt" as among the hindrances which affect and control the sweating of hides. All sweat tanners fully understand that the salt (if the hide is salted or pickled) must be fully soaked out before the hide will sweat; from this, among other circumstances, is deduced the inference that this process is a decomposing one—for so long as the hide is held (cured) by the presence of salt, carbolic acid and other tawing ingredients, the sweats will not operate on it.

So, too, if the hide is covered with grease—as many of our Western and California hides are—it will not sweat evenly, owing to the presence of grease on some portions more than on others. If the hide is covered all over with grease, notwithstanding the action of the soaks and mills, then this grease should be removed. This may be done by an alkali, such as salsoda, soda ash, potash, etc., of commerce; when these cannot be obtained, hard wood ashes, freely used in the soaks, will turn the grease to soft soap, and it will readily wash off in the mills.

NOTE.—For illustration and further description of improved style of cold sweat pits, as at present in use by American tanners, see engravings in latter part of this book.

CHAPTER III.

PREPARATION OF HIDES FOR THE BARK—LIMING.

GETTING READY THE LIMES—THEIR AGE AND STRENGTH—EFFECT OF THE LIME ON THE FIBER—PLUMPING AND BATING—PROF. LUFKIN'S PROCESS—THE "BUFFALO" METHOD—THEIR RESULTS—HANDLING IN THE LIMES.

By far the oldest method of unhairing the hide is known as the "liming process." But old and well established as this system is, there are economical and wasteful methods, both in the use of lime and in working and handling the hides by this process while in the beam house. Without taking up the time of the reader by pointing out the defects, let me as briefly as possible call attention to the most approved methods of using lime for the unhairing of hides.

Lime for the tanner's use should always be unslaked, or in the "stone." When it is supplied fresh from the kiln, as it should be at frequent intervals, it should be kept in a dry and confined apartment, where neither moisture nor air can reach it. This lime should be "slaked" with even more care for the tanner than is exercised by the mason and finisher.

The following description indicates briefly the best method of preparing the stone or unslaked lime for the tanner's use: Have a half hogshead placed near the lime vat it is proposed to replenish; for a pack of 120 to 140 sides throw in, say, one bushel of lime; dampen it by pouring on one or two pails of water, and cover with a thick canvas; a few moments will suffice to absorb the water, and considerable heat will be produced; add water, gently, several times, rather than "drown" the lime by an oversupply at any one time, and be sure and not allow it to "burn" for want of water. The hogshead should be kept covered until the slaking is completed. When the reaction is over the tub may be filled with water and thoroughly stirred; after settling, the liquid or soluble portion should be poured off into the vat, leaving all grit, dirt and unslaked lumps in the hogshead. Nothing but pure lime water should ever be allowed to go into the vat; this will not only render frequent "cleaning out" unnecessary, but will save the edge of the fleshing knife, and shorten the time required in many subsequent operations.

Some tanners prefer old and weak limes; others fresh and strong ones. Where old limes are depended upon, filled with the ammonia of previous decomposing packs, the writer would suggest that damage may result unless care is observed, for this use of old limes involves in a measure the principle of the sweating process. This method, on the whole, may be regarded as only suitable to cold weather; in warm weather it certainly is too dangerous for general adoption.

All tanners will, however, recognize the importance of making a difference between the process for upper, calf, kip, harness and belting leather, and that for sole leather.

For the former kind of stock not less than three or four days should be consumed in the process; this length of time will kill the grease, and fairly plump (distend) the fiber. The effect of the after process of bating is to leave the tissues of the hide relaxed and the fiber elongated—just the condition in which the fiber should be left in order to make tough, flexible leather, for when once the fiber of the hide has been unduly expanded, and the gelatine cells broken or disturbed, they can never again be brought back to the closed and compact condition in which they were found in the natural hide. A microscopic examination of the condition of the fiber of tanned leather will leave no doubt as to the process which has been pursued in the beam house. It is possible so to "starve" the leather in the handlers and the after process by weak decoctions as to break down the distinct membranous cells which hold the gelatine, but usually this is done in the beam house. If, added to the swelling and depleting process in the beam house, a starving process is followed up in the handlers or the layaways, then, instead of plump, well-filled leather, we shall notice a stringy, elongated fiber, consisting for the most part of animal tissues, and, as these do not absorb tannin to the same extent that gelatine does, of course we have no gain, and have, externally, a coarse, broken offal and grain. Indeed, we reproduce the examples which we see daily coming in from small country yards. The waste which this process induces will surely prevent tanners who make such leather from reaping large profits or from competing with those who make the most possible out of the stock which is placed in their hands to treat.

Three or four days in the lime will not improperly fill the hide, and when unhaired it may be speedily reduced to a natural condition. This reduction (depletion) will be well begun by throwing the hides, or sides, into a wheel, and, with a flow of warm water turned on, running for ten minutes. The advantage of warm instead of cold water is very marked, and warm water may, at this stage of the process, be freely used with safety. It is always safe, on hides filled with lime, to use heat to the extent of 100° F., or as hot as can be borne by thrusting in the hand and wrist. This rinsing process will remove the greater portion of the lime, and will ordinarily prepare the hides for the liquor; but some tanners are so particular as to insist on bating still more by the use of hen manure, or some "sour," such as may be prepared with wheat bran or molasses.

For sole leather this latter precaution is quite unnecessary; but for fair or harness leather, and perhaps for calf and kip, where not only a clean but a soft grain is demanded, such extra bating may be justified.

Up to this time the treatment of full grown hides only has been considered, where they have been handled with lime for three or four days. If filled with lime, by being retained in the process for six or eight days, more care and more particular treatment in the bate may be necessary. But the writer would leave all such tanners to work out their salvation as best they may. This is certainly true, that, as a rule, for all upper stock, the hides or skins should go into the handlers as free as possible from lime, while sole leather stock may be trusted to right itself as it passes through the sour liquors of the handlers, provided the fiber has not been unduly strained (expanded) in the beam house.

Thus far the well-known and generally-used methods of unhairing by lime have been considered, but there are other forms in which lime performs a more or less conspicuous part. Referring, in the first instance, to Prof. Lufkin's process, the writer is enabled to state with great particularity its details, through the kindness of Mr. Charles Cooper, of Sparrowbush, N. Y., than whom no man has had greater experience in its use. He has prepared, for several years in succession, not less than 50,000 hides by this process, and with great success, as the superior leather he has produced will attest. His packs were made up of about 50 hides each, either cured, green salted, or dry Buenos Ayres or Rio Grande. The green hides weighed 50 and the dry 20 pounds each. For such a pack he would slake 80 pounds of stone lime, in the manner already indicated, except that the lime is not watered after the slaking process has been finished This leaves the lime of the consistency of a thick paste. While in this state take a small portion and knead thoroughly with 10 pounds each of soda ash and pulverized sulphur; when these three products are well mixed, and while the lime is yet *warm*, turn the whole in together and mix thoroughly; after doing this take lime liquor from the vat and fill the cask, stirring all the while; when completed pour into the vat and thoroughly plunge the whole. No more liquor should be in the vats than sufficient to cover the 100 sides when thrown in. It is desirable to keep the lime thus made up to summer heat by the use of steam, which may be done by inserting a steam pipe while the pack is raised. The handling should be performed once or twice each day if the hides are thrown into the vat in the usual way; but if handled on sticks or reeled, then plunging and more frequent handling will facilitate the operation.

As to the advantages of this process, the writer would say that the sulphur modifies the harshness of the lime and soda ash, and renders them almost as controllable in the hide as soft soap, for, while the hide may remain in the lime for an equal length of time as with the old process, there is not the same swelling, nor the same harshness, and a few minutes of wheeling in warm water will reduce the pelt to almost the consistency obtained in sweat stock. There is no question that it is a good method of unhairing for all kinds of hides or skins, and when a soft and smooth grain is desirable it is a valuable improvement. Of course, it is slightly more expensive than pure lime, and for this reason has not found general favor.

The next innovation upon the old system of liming may be called, for the sake of distinction, the "Buffalo" method.

This method depends largely upon warm, or even hot water, to complete the process. The hide is prepared in the usual way, and is then thrown into a strong lime for eight to ten hours, when it is taken out and immersed in water heated up to 110° F. The warm water soaks, softens and swells the roots of the hair, and very much such a result is obtained as in "scalding" hogs. So little lime really permeates the inner fiber that, after a slight wheeling, the sides may be thrown into cold water and allowed to cool and plump, preparatory to taking their places in the handlers. The process is strongly commended for sole leather, particularly where great firmness of fiber is desired. The tanner who tries it must be satisfied if he gets twenty to thirty sides per man unhaired and fully ready for the liquor, per day.

Besides these, we have many patented methods, both for softening and unhairing, but as they are all founded upon supposed secrets, and have some powerful alkali as their base, the writer will not at present indicate their merits. It will be sufficient to say that, up to this time, they have made no material progress.

It now only remains to describe some of the methods of handling while in the process of liming. The old method is to "throw in" and "haul up" from an ordinary vat—6 feet wide by 8 or 9 feet long. This is so laborious that the sides or hides frequently do not get hauled more than once each twenty-four hours. Among the labor-saving methods in this department the following may be mentioned:

1. "String" the sides or hides by tying pates and shanks together and reel over from one vat to another. [The reel recommended in Chapter VII., on "Handling," and which is illustrated on a subsequent page, will also answer for this purpose.]

2. Hang on poles or sticks by throwing the hides or sides

over, resting on the shoulder—butt and pate down. The lime liquor is plunged and thus a most perfect agitation and handling is effected.

3. Have the lime vats sixteen feet long; place the pack, while yet in whole hides, in one end, resting on each other; each hind shank should be fastened to a rope or strong string five feet long, with a noose at its end, which should be dropped over an iron upright at the corner of the vat. If this noose is no larger than the iron stanchion, of course the top noose will indicate and draw the top hide. One man on each side of the vat will, by the use and guide of these strings, transfer the pack of hides from one end to the other of the vat in a few minutes. This system will always leave one end of the vat unoccupied, and while thus situated it may be "heated up," "plunged" or "strengthened."

There is an advantage in connection with the latter system which cannot be too strongly urged upon tanners. It enables them to lime their hides whole, and to split them after the liming process is complete, and this in turn makes straight lines. For belt or harness leather tanners this is very important. All hides that are split before being limed will be crooked and irregular on the "back line," for the reason that lime takes hold on (contracts) more the thin than the thick portions where the whole hide is equally exposed. The consequence is that the shoulders contract more than the butts. This leads to waste where long straight lines are needed in the cutting, as with belting and harness leather.

CHAPTER IV.

PREPARATION OF HIDES FOR THE BARK—FLESHING AND TRIMMING.

THE BEAM WORK—CLOSE FLESHING, WITHOUT BREAKING THE GLUE CELLS—FLESHING LIME SLAUGHTER STOCK—WORK TO BE DONE BEFORE LIMING—FLESHING SWEAT STOCK—IT SHOULD BE DONE WITH A WORKER—THE GERMAN FLESHER—TRIMMING—CROP LEATHER AND BACKS—ADVANTAGES OF TRIMMING UPPER AS WELL AS SOLE—"ROUNDING"—TRIMMING BEFORE TANNING—BEST METHOD OF UTILIZING THE HEAD, PATE, ETC.

By far too little attention is paid to beam work by American tanners. No amount of labor or care in the after process can atone for neglect in this department. The flesh should all be removed, and the natural structure of the hide should not be disturbed or even touched with the edge of the flesher. The difficulty of accomplishing this with the careless, unappreciated and unrequited labor at present employed in this service, is great; but when tanners come to understand that both national and international tastes and wants demand that this work be properly performed, there will be no difficulty in securing the necessary reform. It is estimated that the additional fleshing will cost, in labor, three cents per side with lime stock and two cents per side for sweat stock, while it will deprive the side of about one-half a pound of fleshy, fibrous matter, when tanned. This, then, is the sacrifice tanners are required to make in order to meet the foreign demand and taste, as well as give better satisfaction to consumers at home.

If the question was between removing the flesh and cutting the cells that contain the gelatine on the one hand, or leaving all the flesh on and retaining the structure of the hide intact, tanners should of course prefer the latter; but there is no such alternative. The flesh may be removed without injury to the structure of the hide, and this should be done by all tanners who pretend to commendable workmanship. A clean flesh is desired not more by our own manufacturers than it is demanded by those of Europe, and both should be gratified. If persistent and dogmatic statement will accomplish reform in this important department of the tanner's art, the writer proposes to use both in the way of entreaty and admonition.

The fleshing of "lime slaughter stock" and of "dry hide sweat stock" are quite different operations, and must be considered separately.

Our practice in regard to the first class is to throw the hides into the soaks for a day or two—just long enough to cleanse all the blood, salt and other impurities from them—then draw from the soaks and split, and throw into the lime, paying no attention to the flesh until they come on the beam, after passing through the lime. The hide is plumped by the action of the lime, and the theory is that, when the flesh is thus swollen and raised, it may be removed with less danger than if done before the hide was plumped. This is undoubtedly true, if only the same labor and skill is to be employed in each case. But, on the theory of accomplishing fully what we concede to be desirable, the "meat" must be removed from the flesh before the hides go into the lime, and the meat should be "worked off," and not "cut off." The "cutting off" implies skilled labor, but the "working off" is more economically done by willing, unskilled labor. When the meat on the hide (which should have been left on the carcass) has been removed, two things will be accomplished: 1st. The action of the worker will have, to some extent, broken (distended) the nerve of the hide; and, 2d. All obstruction to the uniform liming of the hide will be removed, for patches of meat on the flesh side obstruct the action of the lime, as all good tanners concede.

When it becomes fully understood among tanners that all the flesh from slaughter limed stock must be removed not only in the beam house, but while the hides are whole and before they go into the lime, it is not unreasonable to suppose that some mechanical contrivance will be devised to accomplish the work. At present we are compelled to overcome this obstacle by manual labor, and hard, disagreeable work it is. But the actual money cost is not more than six cents per hide, and as this labor, thoroughly performed, will render the after beam work much easier, the writer ventures to say that no tanner making this stock can long resist the innovation. Many of our upper and harness leather tanners do at present remove the meat before liming, but the practice is not general. In another chapter the reason for liming the hide whole has been explained, and that subject need not here be entered into.

The fleshing of sweat stock is quite different from that necessary where lime is used. The beam hand is always cutting or working on a soft, pulpy substance, and should never use the edge of the flesher except to trim the edges or cut the filmy portions attaching to the flesh. All extraneous flesh and hair may be readily worked off from sweat stock for the price already indicated, by a fine tooth worker; but these teeth, if long and sharp, are likely to cut the outer cells of the fiber. It would be better to remove the flesh with a smooth-edged worker, which may be done at a cost not to exceed two cents per side. The saving in the amount of flesh and tissue removed, and the damage avoided from the cutting of the fiber, will amply repay the extra cost. There are two machines, either now offering on the market or soon to be introduced, one of which proposes to remove the flesh while in the beam house—the power beam worker, of Mr. Lampert, of Rochester, N. Y.—and the other is intended to shave or cut the flesh off after it is tanned—the new buffing and whitening machine, introduced by Mr. Caller, of Salem, Mass. But until one or both of these machines have demonstrated the success which is claimed for them tanners had better work off the flesh, as already indicated.

The usual flesher and half round beam are too familiar to the tanner to require notice in this connection, but within a few years the French and German beam knife has been introduced into this country and received with general favor. This knife is about one-third longer than ours, and is not more than two inches wide; the material is the best steel, and the knife is not more than a quarter of an inch in thickness on the rib or center. The blade is so supple that the handles can almost be brought together. It is claimed for this knife that, by bending round the convex form of the beam, it makes a flatter cut on the flesh of the side, less concave than a stiff, straight-edged flesher cutting on an oval or convex surface. To this extent the new knife certainly does present advantages, and may be safely trusted to do good work in skillful hands.

In the year 1838, a tanner in the State of New York, in obedience to the suggestion and order of a manufacturer of boots and shoes, commenced to "crop" his sole leather. The full extent of the demand at first was 100 sides per week. This demand grew, as other manufacturers came to see the economy of the practice, until 2,000 sides per week were required in 1844. And now, in 1876, probably not less than 1,000,000 sides of "union crop and backs" are manufactured yearly in the United States, and sold both in this country and Europe. Within a year or two the "crop" form of trimming has largely given way to the "back" form. The latter differs from the former only in the removal of the pate, at the point where the throat is usually cut, which makes the shape of the side more compact, and in harmony with the "bend" leather of Great Britain. There must be some advantages in this method of trimming, or otherwise the trade would not have grown so rapidly, and, to acknowledge this, is to concede that still other improvements may be made. Let us inquire what they are.

The tanners of Great Britain, who certainly study the economies of their trade more thoroughly than the same class in any other country, not only trim the bellies, but also the shoulders, from their butts and bends. The following reasons may be assigned for their practice:

1st. The shoulders and offal are much thinner than the butts, and therefore tan in shorter time.

2d. The offal being used where a tough fiber is required, slack, or, at most, a full tannage is all that is required.

3d. The boot and shoe manufacturers, not only of Europe but of America, have so classified their work that those who use butt leather largely do not require so much "inner soling" and "welting" as would come from the bellies and shoulders of the hides which would give them just the description of sole leather required.

But the upper leather tanners of Great Britain also trim (round) their upper leather, notably so their East India kips. The good sense displayed in this process should be adopted by the tanners in this country, particularly on all East India,

Russian or native murrain kips. The bellies and shoulders of these, when finished on the grain, make a most serviceable leather for women's and children's shoes, while the butts, finished on the flesh, answer for a stout boot or shoe. This method of rounding light upper sides and kips recalls the economy of our fathers, when it was customary to make "magpie" leather—that is, grain and wax finished on the same piece. The farmer had his single hide, taken off in the fall, tanned by the halves, and finished "magpie." This gave him the thin bellies and offal, grain blacked for women's and children's shoes, and the thicker portions, "waxed," for wear by himself and older sons. One of these days we shall imitate the economy of our fathers, and treat the whole shoe and leather consuming people as one family, whom it is our duty to supply with leather fabrics upon the most economic plan, and when this time arrives we shall find ourselves imitating very closely the habits and practices of the leather and shoe manufacturers of Great Britain, for they are, no doubt, far in advance of us in all these respects. It is probably true that the population of Great Britain is better and more economically shod than any people in the world. A portion of the economy is due, however, to iron rather than leather. Much of the economy here conceded arises from their method of trimming and rounding their hides and skins. That which belongs to the glue maker never goes beyond the beam house in any tannery in Europe, and this includes the hide from the pates, heads, shanks and tails. In treating of the "savings of the tan yard" their methods of utilizing these as well as the hair and other serviceable offal will be explained.

Our best city custom harness makers can scarcely realize the fact that, until within a few years, they have been compelled to buy the offal of all the sides they cut. Now they buy only the backs, or just such portions as they need, while the trunk, collar and strap leather manufacturers can get their supply from the bellies of these sides. This economical arrangement we borrow from Great Britain.

There is one advance, however, in the economies of leather trimming and cutting which owes its origin to Lynn, Mass., viz., the cutting of soles, and sorting them to suit each special manufacture. This gives each what he requires, and only that. Upper and calf fronts are furnished in a similar way to manufacturers in all Southern Europe, but, so far as the writer has observed, the practice of furnishing soles to large manufacturers is confined to this country.

The trimming of hides used for belts, bags, and more recently for enameled leather for carriages and furniture covering, should by this time be familiar to all. The practice, however, marks an advance in the progress of the art which gives promise of infinite extension and profit.

In closing this chapter the writer wishes to offer a suggestion. The head of the ox or the cow is now skinned while yet attached to the body, and, as it is done by the usual skinner, who comprehends that the hide is more valuable by the pound than the coarse meat, he leaves as much of the meat as possible on the skin rather than the head, and this policy will probably continue until the head is severed from the body while the skin remains on, and is in this condition handed over to men or women whose special training will induce them to skin the cheeks and throat together in one piece, leaving the "pate" proper to be skinned and wholly handed over to the glue maker. If this is carefully done, the skin of the cheeks and throat will give four or five feet of fine grain leather, besides two or three pounds of glue stock, for this head piece should be split while in the lime, and only the grain sent into the yard to tan, while the split portion should go for glue. The nose and lips should be cut off while fresh, and handed over to the sausage makers, who so well understand how to scald, pickle and prepare these delicate morsels for human food. The meat should be carefully cut from the head while fresh, and made into sausages. The bone, including the pith of the horn, should be crushed or ground into a fertilizer, and the horns handed over to the button or comb maker.

Many of these processes are now adopted in the final disposition, but all are retarded, and the result injured, by the mistake of skinning the hide from the head while attached to the carcass. Let the tanners consent or insist on this separation of the head when they buy their hides, and we may depend upon immediately taking one step forward in the direction of a more economical method.

CHAPTER V.

GRINDING BARK—BARK MILLS.

THE INEXPENSIVE AND ABUNDANT POWER OF SOLE LEATHER TANNERIES—GRINDING BARK FINE AND UNIFORM—USEFULNESS OF SCREENING THE GROUND BARK—THE DOUBLE-GRINDING MILL—ITS EFFECTIVENESS WITH DAMP BARK—THE ALLENTOWN MILL—ADVANTAGE OF A STRONG MILL AND WEAK COUPLER—THE SAW CUTTING MILL—A BARK CRUSHING MACHINE—THE PROPER SPEED AT WHICH MILLS SHOULD BE RUN—TANNIN LEFT IN THE BARK.

It will be conceded by all practical tanners that the preparation and proper grinding and leaching of bark stands at the head of all the economies of the tanner's art. The very large number of imaginary and real improvements made in this department are so many concessions to the importance which is attached to it by tanners and inventors. Patents almost without number have been obtained on bark mills and their attachments, and on leaches and their various methods of heating and handling. Life is too short and space too limited to review all these improvements and patent claims. It will be sufficient for the present purpose to indicate the methods and processes which have gained the most general adoption in our largest and best constructed tanneries.

The successful burning of wet spent tan in detached furnaces has, more than any one or all causes combined, contributed to a reform in the grinding and leaching of bark. Whether to furnish the motive power to grind the bark, or the heat to extract the strength, wet spent tan has afforded a substitute for water power and wood or coal fuel, and has proved so complete a substitute that all previous expedients have been abandoned. So absolutely inexpensive is this material that power and heat may be used without stint or limit in the manipulations of all our modern sole leather tanneries. "Sole" leather tanneries are particularized because upper leather tanners usually are in short supply of refuse tan to do all this work, and they even now find it profitable to dry or partially dry their tan, with all the improvements for burning wet spent tan open to their use, thus proving that "water" does not increase the heat-giving properties of tan, but diminishes them, notwithstanding the learned opinions of experts to the contrary.

The fact then stands conceded that the wet spent tan from an ordinary sole leather tannery will give ample power to grind all the bark and heat all the liquor required. If any further saving of labor can be suggested by the use of more power and more heat, they can readily be obtained by the surplus tan now thrown away. Therefore we go into the economic consideration of this question with an absolute mechanical power without limit. The tanner may grind to any degree of fineness; he may screen and return to the mill any portion of his bark; he may convey up or down, or laterally, to any required distance, either his dry or spent tan. With such power, if he fail to extract the whole strength, and do it in the most acceptable and satisfactory manner, he has no one to blame but himself.

The importance of grinding bark fine, and yet uniformly so, without dust on the one hand or large coarse lumps on the other, is conceded to be very great. Dust obstructs, and coarse lumps prolong the process of leaching. Imperfect grinding does more: it necessitates the use of extreme heat to extract, whereas if the bark is uniformly and properly ground very little artificial heat will suffice in winter, and no more heat than that of water at the ordinary temperature will be required to extract the tannin in summer.

For more than forty years tanners have been seeking the best bark mill. In this respect they are like the farmers who seek the best plow. Both bark mills and plows are comparative, not positive, in their character and attainments. Before determining in any given case just what mill should be adopted the circumstances should be known. If the tanner has purely hemlock bark or purely oak bark, and, moreover, if these barks vary in condition, being sometimes damp or wet and at other times dry, these conditions should be known before an intelligent opinion can be given as to his wants.

For the grinding of all kinds of barks, with the most variable conditions, I judge that the "double grinder," formerly known as the "Starbuck" or "Troy" mill is preferable. The patent on this kind of mill has long since run out, and it is now manufactured at various tanning centers, but while all are essentially the same in construction, their "fitting up" makes the difference between a serviceable or worthless mill. A double grinding mill, carefully fitted, will grind damp or wet bark with more success than any other mill, and for the reason that its grinding surfaces are open—set with obtuse angles—rather than close, with sharp, acute angles. There is no reason why this mill may not always hold a respectable position among tanners when proper care is bestowed upon the castings and fittings.

The most artistic, and, on the whole, the most economical mill yet presented for our adoption is the mill made by

Messrs. Wm. F. Mosser & Co., of Allentown, Pa. This mill, at first, costs about $70 to $80, but may be renewed for from $6 to $10 for an indefinite period. The principle on which it is constructed must be approved by all, and cannot fail to be generally adopted. Its large cost lies mainly in the original "fitting up." The shaft or spindle is made of wrought iron (or should be so made), and is carefully turned at its bearings. The bowl and curb are also turned true, and segments, either of cast iron or steel, are bolted on these turned surfaces. This form of construction gives a true and adjustable grinding surface which leaves nothing to desire. These segment surfaces can be either sharpened with a cold chisel, or a "saw gummer" run by power, or may be renewed with new iron or steel segments, for the inconsiderable sum above named. The original cost of a mill then is of small consequence. It is the cost of frequent renewals that makes the expense of grinding with such mills so great. By far the largest replacement is made necessary by breakage, and not by wear, and hence the importance in saving the original and more costly structure. This is done in the case of this mill by a safety coupling, which renders breaking by the usual casualties impossible.

So confident was the possessor of one of these mills that no accident would result by throwing in iron, that, in my presence, and against my earnest protest, he threw in an iron bolt with the bark. In a moment a slight "click" was heard. The mill stopped, but the driving shaft went on. The coupling only had broken (this being the weakest point). This coupling costs only seventy-five cents. In ten minutes the step of the mill was lifted into a new coupling, the iron picked out, and the mill started as usual.

If this safety coupling can always be relied upon, then, beyond the original cost, a new mill can at all times be procured of iron for $6, or of steel for $10, and this replacement is made necessary only after repeated sharpening of the grinding surfaces by the ready means already indicated. Two hours will suffice for removing, grinding and replacing one set of these segment surfaces, and so true do the mills run that they will "crack corn" successfully, or would be acceptable coffee grinders for a large army of men.

There is one thing, however, that these mills will not do. They cannot successfully grind damp or wet bark, and it may be doubted whether, in the nature of iron-mill construction, a close and fine bark grinder can be made to perform this difficult service. This is a problem at which the owners of the mill are now at work. Whether they succeed or not, it is certain they now have the best, and, all things considered, the cheapest bark mill known in this country, and for the grinding of dry bark it leaves nothing to desire.

It has always been difficult in practice to secure absolute uniformity in grinding, even by the newest and best iron mills. The defects of grinding can be overcome by the erection of a screen or wire sieve, stretched over a revolving skeleton frame of wood. The meshes should be three-sixteenths of an inch square, and, thus constructed, may be relied on to "bolt out" all the coarse particles, and send them back to the mill to be reground. This inexpensive contrivance will cover the defect of leaving too coarse particles in the ground bark, whether these come from an imperfect mill or an old, worn-out one, and no tanner can safely do without it. With the "Allentown mill" the screen will have less to do, but even with this uniform grinder a screen is useful. With a properly constructed screen, almost any mill, whatever its condition, may be trusted, as its defects can never reach the leach.

The limited service now performed by this screen only partially indicates its possible usefulness. There is no reason why all the bark from the mill may not go into a tight inclosure supplied with wire "bolts" or screens that will make as many classifications of the ground bark as, in the judgment of the tanner, are demanded. The dust should be sent to a leach made specially to extract its strength. This may be a revolving square or round leach, with inside projections to agitate the contents, or may be a leach with a round bottom, with a paddle wheel revolving on and in its top surface, for the purpose of keeping for a time the dust floating in the weak liquor or water. This bark dust so readily yields up its tannin that almost any contact with warm water will suffice to denude it of its tanning properties. When the dust is all separated from the remainder of the bark, the leaching will proceed rapidly, as the circulation is less obstructed.

Of course the ground bark of this screening process can be graded to suit either the leach or the layaways. The latter service requires a much coarser bark than the former, and, by careful screening, bark of any particular size which may be needed can be obtained.

The question is often mooted as to the degree of fineness to which bark may be reduced for advantageous leaching. The answer is, as fine as the "percolation" or "press system" of leaching will permit. If all bark could be reduced to a uniform size, "buckwheat" or "cracked corn" would nearly represent the condition most favorable for the ready and economic extraction of the tannin from the bark.

Within a few months another example of "saw grinding" or cutting of bark for tanners' use has come to my attention. The work performed by this machine is admirable, and if done with the ease and rapidity stated by those interested, then it promises to divide with the best bark grinder the patronage of our tanners. The promoters of this sawing machine, however, seem to think that their machine is new. In construction it may be, but not in principle.

The numerous other mills made are not here mentioned, not because they have no merit, but mainly because their advantages are so varied and disproportionate to their number that it would take up too much space and time for their presentation. Most of the bark mills which have considerable merit, and have been largely introduced to the trade, have had their specially commendable points presented to our tanners through the columns of the *Shoe and Leather Reporter* during several years past. But the writer has these suggestions to make to bark mill manufacturers: Take great pains in fitting up your mills; make your grinding surfaces replaceable, and have them run true, so that what bark goes through is not ground to a powder, but in uniform particles; adopt a safety coupling, so that a valuable and even a high cost mill can be indulged in by all tanners at an inconsiderable aggregate outlay.

On this subject it now only remains to call attention to a "bark crushing" and "extracting" machine, which came to the attention of tanners a few years ago, but has now been forgotten by its failure, and is here mentioned to call attention to the fact that bark crushing, like bark sawing, has been tried. This machine was a most ponderous affair. It consisted of a series of metal rollers, through which the unground bark (bark in the leaf) was passed after having been soaked in hot water for an hour or more. It was claimed for this process of squeezing that nothing but the fine "salts" would come out, leaving all the resin and much of the coloring matter behind. The report which was made on the practical working of this costly experiment was to the effect that the bark (hemlock) was thoroughly crushed, and a portion of the tannin was squeezed out, but that much remained. It was conceded that the bark was most admirably prepared for leaching, being left in the form of a "pulp." But the machine was expensive, costing fully $1,000, besides requiring 30-horse power to drive it, and, while interesting as an experiment, was a failure for practical work. When last heard from this machine belonged to a corporation or association, and was started in Tennessee to make oak extract.

It is quite common for patentees and even tanners to over-estimate the amount of bark ground per day or hour by their mills. It is a good mill that will average one cord per hour, although we hear reports of the grinding of one and a half and even two cords per hour. But bark ground at the rate of one cord per hour and well done is far more profitable than more rapid work. The writer has seen a mill grind one cord in fourteen minutes. The mill was, however, "geared up" to 280 revolutions per minute, and ground the bark very coarse at that.

This leads me to say that a slow motion is desirable. It is questionable whether a motion of over eighty revolutions per minute is either profitable or effective; certainly the benefits can in no wise overbalance the defects in grinding and danger of fire arising from excessive friction. A quick motion has the effect to "throw up" and "back" the bark, rather than to take it in and pass it through the grinding surfaces, as a slower motion will. The best experience has demonstrated that eighty revolutions on a small and seventy revolutions on a large-size mill is the proper motion. Any faster motion, besides the loss of power involved, will not proportionately increase the result, and will, besides, greatly increase the fire risk.

The time may come when hemlock and oak bark will become so scarce and dear as to necessitate other means than grinding and leaching for getting the strength out. It is now est mated that from 7 to 10 per cent. of the strength is left in, and it has been doubted whether any of our present methods will take out that small remainder profitably. The English tanner, who pays from £5 to £7 per ton for his bark, would pitch the leach over once or twice, releaching each time until the last particle of strength was out. The German tanner would use his bark on his layaways for two or three months, and then take the last strength out by leaching. All these processes are so different from ours that we cannot avail ourselves of their tedious methods. The time may come when it will pay to pass the partially spent tan through metal rollers, thus breaking and crushing the unspent portions. So far as the power is concerned, this could be afforded now; but the labor of passing the tan out of the leach and through these rollers, and then again back into the leach, would cost more than the small percentage of strength gained would be worth.

Some illustrations of bark mills, and sectional parts of mills, with further description of their working, will be found in subsequent pages.

CHAPTER VI.

LEACHING.

TANNIN VS. RESINOUS AND COLORING MATTER—TANNIN REQUIRED TO MAKE GOOD WEIGHT—EXTREME HEAT IN LEACHING INJURIOUS—FLOODING THE BARK—THE PRESS LEACH—HEAT TO BE APPLIED ONLY TO THE WEAKEST LEACH—CONSTRUCTION OF LEACHES—CLAY AND LOAM PACKING FOR THE SIDES AND BOTTOMS—WORKING THE PRESS LEACH—THE SPRINKLER LEACH—REVOLVING DETACHED LEACH.

The full and perfect extraction of all the tannin from the bark is not only desirable, but is of primary importance; it is equally essential, however, that this subtle elixir should be extracted without deterioration or injury. It is found in practice not at all difficult to wash out all the extractive or soluble matter from bark, but to separate and take out the greatest amount of tannin, leaving the largest portion of coloring and resinous matter behind, is quite a different thing, and one which has taxed the efforts of our best tanners. The system of leaching which will best enable the tanner to control and separate these qualities is, in my judgment, the one to be most commended. In fairness it ought to be said, however, that there is much difference of opinion on this subject among some of our largest, and, financially, the most successful of our tanners.

Their theory and practice must proceed on the assumption that coloring matter will give weight, when incorporated with the hide, and that the resinous matter will in some mysterious way attach itself to the leather, defying the action of the scrubber to wash it out. This, to the writer, is a dangerous heresy, and should be rejected unless there are facts and considerations favoring it which have never been presented to the public, for to it may be justly charged all the defects in color which for so many years baffled our hopes of a successful foreign market. It has also caused other defects, both of finish and quality, which it will be more appropriate to consider in another place.

The commencement and successful continuance of the "union" sole leather trade, both in "crop" and "sides," has demonstrated that it is possible to make a light, fair color while using hemlock bark. With such experience as the manufacture of this description of leather furnishes, the writer is prepared to defend the position here assumed, viz., that tannin, and not coloring or resinous matter, enters the fiber and gives weight. All the illustrations or seeming proofs to the contrary are drawn from heavy sole leather tannages. There is an inherent inequality in the conditions of the two kinds of tannage which will fully account for the disparity in weight made under each, quite independent of the conditions we are combatting. It would be unfair for me to assume the correctness of my theory from the fact that a given quantity of bark will make more light colored leather than dark, for bark will go further, as all know, in light weight tannages than in heavy. This is illustrated by the well-known fact that one ton of hemlock bark will tan 300 pounds of upper leather, while it will only tan 200 pounds of sole leather. It would be as unfair for me to assume that this difference was creditable to the absence of the extra coloring and resinous matter, as it would be for those who differ from me to assume the opposite conclusion.

The success and value of any system of leaching must depend upon the intrinsic or serviceable *value* of the liquor or extract obtained. If liquor obtained from the bark without heat will make leather that will bring in the market one cent per pound more than leather made with an extract obtained by extreme heat, then it is clear that a cold or more moderate process is preferable to the hot water or steam process, provided all the other conditions are equal.

The "union crop" leather tanners have learned to comprehend the value of moderate instead of extreme heat, and when better methods of grinding and screening the bark shall be appreciated at their full value, even less heat than at present will be employed by them; the more nearly summer heat (60°) is adhered to on the head leach, the more modified and controllable will be the color. This view of leaching was maintained by Mr. James Clewer* during all his American experience, and it must be conceded that he produced as satisfactory results as any tanner, either before or since his time. He frequently demonstrated his ability to make 190 to 200 pounds of leather with one ton or cord of bark, and he never used more than summer heat on his head leach except in winter.

There have been during the past forty or fifty years a great variety and many forms of leaches and leaching processes. Previous to this period, viz., about 1820, leaches were not used to any considerable extent for new bark. A single vat was set apart into which all the old layaway bark was cast for final washing, but this was the extent of the leaching process; and if the writer's observation and information can be relied upon this sytem prevails now in most of the coun-

*An English tanner who came to this country about the year 1829. He was a practical workman, and had as thorough an understanding of the true principles of good tanning as almost any one in the trade at that time, either in this country or Great Britain. His work and experiments among our tanners were of lasting value in materially advancing the trade in many particulars.

tries in Southern Europe. No country now so universally employs leaches as our own ; and, therefore, we can draw no lessons from their larger experience, as in many of the other departments of our leather manufacture.

There are in this country three distinct systems or methods employed for leaching bark, but a much greater variety of forms. For the sake of designation we may call the first a " douse," the second a " press," and the third a " sprinkler " leach. The douse leach is the oldest known method. It is the one used from 1824 to 1832 by the hemlock tanners who made their way into Greene, Delaware and Schoharie counties, in the State of New York, at that early period of our tanning history. Usually one leach was placed on the top of another. The top leach had a heater of some form contained within it, by which the liquor and bark were " heated up," even to the boiling point, and the liquor was finally dropped down on a leach below, usually also containing bark, but not always, for, when stationary, it was found difficult to so run them as to wash properly the weaker bark ; usually the upper and lower leaches, however, were filled at the same time, and worked together as one leach. Any leach, or system of leaching, may be said to belong to this class that floods the bark and allows the whole bulk to remain saturated and stationary. Even where the leaches are pumped over from one to the other, if the contents are allowed to stand, and the liquor is then drawn off in bulk, the leach belongs to this class.

The "press" leach was introduced into this country by Mr. James Clewer, about 1830, and the method was improved, by his own suggestions, from some leaches which he had worked in England. His system washes out the extractive matter, including the tannin, "by column," as contradistinguished from the "sprinkling" or percolating method introduced by

Allen & Warren, which is the third distinct process, as is claimed; and yet it will be seen by close observation that these last two methods resemble each other very much in principle, since both move constantly from the weaker to the stronger bark. The one passes by column and the other by percolation, and that is the only difference.

Keeping in view the three general divisions or classifications of leaches, let us go back and consider them in their order. The douse leach may be of any size or construction. It may be round or square. It may be filled with bark and the liquor or water run on, or the bark may be run on with the liquor (floated from the mill). This leach may be heated up in any way, either by inserting steam under the false bottom, or the liquor or water may be heated before it is run on. This process of flooding and "running up" and "off," either with or without heating, is even now much practiced, and, but for the fact that it tends to dilute the liquors in the yard, is very satisfactory. But this dilution, or want of concentration, is an evil which all tanners deplore and desire to avoid in heavy tannages. For upper, harness, and calf, it will do better service; but, for reasons presently to be stated, there is less thoroughness in the leaching than by either of the other methods, for it must be evident that the discarded bark must have retained the strength of the last liquor that has passed off, and for this reason can never be *perfectly* denuded of its tannin.

The usual or best form of constructing and working the press leach may be stated as follows: The number of leaches in a set should correspond to the number of days in the week, or the number of full leaches required, multiplied by the number of days in the week, thus, 6, 12, 18 or 24. According to my judgment, six leaches in a set are always better than any other number. By this plan one new leach is filled in each set each day of the week, and if more than the capacity of one leach is required for the day's work, then the sets of leaches should be multiplied, rather than to break into the system as here contemplated. This will give at least five days for the leaching of all the bark, and this time is ample and more than enough. Too many tanners, however, when crowded, break into their plan, and attempt to force two leaches a day from a single set, which creates confusion and leads to waste.

Under no circumstances is heat applied to any other than the back or weakest leach. This will bring the strong liquor of the set on the head leach comparatively cold, or at most at summer heat. Such a course will bring all the liquors passing through the yard into a proper condition to go upon the leather without coolers or waiting. It will also do much more; it will leave behind much of the coloring and resinous matter, and send forward a pure tan liquor, free from all sediment and impurities.

The form may be square or round, depending on the situation of the ground where placed. Where it is possible, this kind of leach should be set in the ground on a level, and well filled in with clay or loam. Exceptionally the writer has seen them set on a side hill—one raised six or eight inches above the other—but it is safe to say that, when placed in sets of not more than six, there will be no difficulty in working the press. When placed in the ground and filled in with loam, they should be, in size, 6 by 6, 8 by 8, 10 by 10, 12 by 12 feet, etc., their capacity being regulated by the size of the vats in the yard.

It may be well here to mention a fact that is not generally known, viz., that loam is quite as good as clay for packing for the sides and bottoms of the leach, if only properly prepared. The preparation consists in mixing the earth with water to the consistency of thin mortar in a vessel before it is poured in between the vats or leaches. If so mixed and run in, a solid sandstone formation will result. To be convinced of the efficiency of this method of filling in between leaches, try this experiment: Take an ordinary pail and fill it with loam; while doing so pour in water, so that when filled there will be a thick earthy substance; stir up well and let it stand for one day. The water will all be on the top, and the earthy portion will have gone to the bottom, each particle in its order of specific gravity. These particles will so compactly adjust themselves as to form a solid stone. Indeed, geology informs us that this is the way that sandstone is formed in the earth. You may bore any number of holes in the bottom of the pail, but no water can pass. The water will in time evaporate from the top, but cannot escape otherwise. Unless great care is observed this concrete will force the sides of the vats or leaches in; or, still worse, raise them from their beds, by working under and pressing them up. To avoid this, water should be run in to fill the vats as fast as the concrete fills the outer sides or passages around and between. It is never safe to trust to studding to keep the new and empty vats in place. Mr. James Clewer used this loam filling with such effect as to render corking unnecessary. But he was careful to have tight joints before battening.

The "covering" of liquor of one leach should in quantity supply one or two liquors for the layaway. There should be no fraction or portion left over. The leaches should be made of hemlock or pine plank, not less than two inches thick, battoned together and corked. Two days' work will make a leach 10 by 10, and one day will cork it. This is additional to the preparation of the ground and the after filling with loam.

Round leaches are recommended when it is inconvenient to

place them under ground. These are more expensive in the first cost, and will la-t hardly more than four years if made of hemlock, and five or six if made of pine; but by reason of the more rapid action of the sprinkling process it is claimed that one-quarter of the leach capacity under that system will give the full result of the less active press system, and this statement is probably well founded.

Round leaches may be used on the press principle, but the square leach cannot very well be used by the Allen & Warren, or sprinkling process, although some attempts of this kind, moderately successful, have come to my knowledge. In this general statement it is supposed that the reader understands the method of making round as well as square leaches, and of connecting them, both from above and below, with the junk yard, waste conductors, etc.; therefore, mention is made in detail only of such points in regard to the connections of the press leach as are peculiar to its system of working.

Let us suppose six leaches all placed on a level, each of them having a false bottom, and constructed in the usual way. Each of these leaches have a tight conductor leading from under the false bottom up to the top of the leach—so tight as to prevent the passage of any liquor except such as has been through the circuit and down under the false bottom. If these tight tubes or conductors are placed in opposite or alternate corners of the leach, it will often prevent currents, which are likely to form if they are otherwise placed. The effect of placing a tight tube in this position is of course to cause that tube to be always filled with such liquor only as has passed through all the bark, and is equal in strength to that standing under the false bottom; it will, besides, stand on a level with the liquor in the leach, and will overflow when it reaches up to the opening which leads to the next adjoining leach. These openings should be about 8 to 10 inches from the top in each leach, and about 2 inches in diameter; if the openings are oval or oblong, all the better. To prevent currents forming it is usual to cover the bark with boards, so that the liquor will spread over the whole surface and press from the top. But if the system is worked uniformly there will be no difficulty from currents forming. It is always safe, when the leach is filled with new bark, to carefully level it off and cover the whole surface with boards perforated with inch holes, the whole to be battened down, so as to hold the bark in place. These board coverings answer the double purpose of holding the dry bark from rising (floating) and spreading the weak liquor gently over the whole surface.

Warm water or liquor is more expanded and consequently lighter than cold. Strong liquor is heavier than weak liquor. Now, if we put the two light conditions together, viz., hot and weak, and place them on top of the heavy and cold liquor, they will remain separate for all time, or so long as these unequal conditions are maintained. A simple experiment with water and tan liquor in a tumbler will demonstrate this practical result. Weak tan liquor will stand all day on the top of strong and heavy liquor, if not agitated. Now, if to the weak liquor we add heat, the separation will be still more marked. A tumbler half filled with weak and warm liquor may be forced out of the top by gently inserting strong and cold liquor underneath by means of a pipe. The action of these bodies, if the experiment is carefully made, will satisfy any one that the system of press leaching can be carried on without mixing the liquors on their passage, if the system here recommended is followed.

The six leaches should at all times be, equally, from two-thirds to three-quarters full of bark, and stand covered with liquor varying in strength with the strength of bark in each. Let us suppose that the head leach has just been filled with fresh ground bark; the last or back leach is then full of spent tan ready to pitch, and the intermediate four leaches are divided both as to strength of liquor and age of bark from these two points. No liquor is sent into the yard except from the top of the head leach, and where very strong liquors are needed only *one liquor is so sent from each head leach*, so that the accumulated strength of all the bark in one leach is concentrated in this one liquor. If a less degree of strength is required, then two runs may be taken off, and in exceptional cases even three or four. The liquor thus sent into the yard is not returned until all the strength is taken from it, and is then either sent off into the stream as worthless or sent again to the back leach after passing through the heater, and heating up to 100° or 120° of heat. The spent liquor, with its accumulated acid, should *go back* again into the leaches for further use if the tanner is making sole leather, and should *go off* into the stream if light leather, such as calf, kip, or even harness, is to be tanned.

If this system is carried out each layer or particle of bark must be washed as many times as the whole bulk exceeds the number of these particles or layers. Suppose the leach to be six feet deep, and to be filled four feet with bark. If the covering of liquor is represented by the remaining two feet (practically it will be more than this), then each layer of an inch will have these two feet of water pass through its particles by corresponding portions of this liquor. So that, in fact, the number of times each particle of bark is washed by the liquor on its passage is almost infinite. This liquor gathers strength all the way on its passage, both theoretically and practically, as tanners may demonstrate by taking samples from any stage of the progress. If the press is properly worked the result will be the same as if six leaches were placed on the top of each other, and the whole amount of liquor percolated through all, or the same as if the bark was leached in one leach 36 feet deep.

One of the advantages of leaching by column in this way, rather than by percolation or flooding, is that currents are not so likely to form, and the fine bark dust is not forced to the bottom to stop the free passage of the liquor. Currents are more liable to form with any other known mode of passing liquor through bark than by this method.

While the writer regards six as the proper number of leaches to connect in one set, as many as twenty-four are sometimes so connected; but when so many are connected it necessitates the use of pumps to aid the press by more or less forcing of the liquors. Covering so much space, with so small a variation in the weight of liquor in the leaches, necessarily makes the flow sluggish.

In regard to the Allen & Warren leach, perhaps the fairest way to present its merits would be to give extensive extracts from the patentees' published circulars. But this would carry me far beyond my purpose, which is to indicate in the fewest words the conceded merits of the various methods of leaching. In this view, it would be unfair to hold this patent accountable for all the damage done by those who use it. There is no reason why hot liquor or water should be percolated through newly-ground bark, as is the custom of many tanners who use this leach. Nor is there any necessity for running into the yard the fine dust and sediment which naturally runs off with the liquor through the open lattice bottom. Both of these practices are rather an abuse of the system, which may be avoided. Cold liquor can be used on the new bark by this process as well as any other. This should, to follow the principle of the press leach, be the strongest liquor at the disposal of the tanner. If, for instance, a liquor of 16° by the barkometer is run on through the sprinkler, it should turn off a liquor of double that strength.

The patentees of this form of leach have laid down certain formulas and claim results which are not usually attained by tanners, and which the writer is inclined to think are not realized in practice. Their claims, then, which are based on superior results over any other system, are not realized No tanner can get more strength than all there is, and there are several methods which practically accomplish this.

This sprinkler leach does concentrate liquor more than any other system, as it is worked, *i. e.*, a given or limited quantity of water will carry off more tannin by this process than by any other. In still other words, the tanner can control his strength better by this method than by any other. For instance, the first barrel or hogshead of liquor which percolates through may stand as high as 32° to 36°, and then the strength begins to decrease, so that when the bulk of the leach has run off the whole aggregate strength may not be over 14° to 16°. The tanner may stop at any point during this percolating process, and thus secure just as little bulk and just as much strength as he desires. There are many forms of this leach, and as many methods of economically working them. The newly patented form (the McKenzie patent) now so successfully making extract at Vandalia, N. Y., is one which is claimed to be very economical. The leaches in this case are round, and only about three feet deep, and are movable. The whole leach, when filled with newly ground bark, is placed under two others of like size. The percolation continues through each until the new bark is reached. It is claimed that these filled leaches are handled by machinery so economically that one man, or even a boy, can empty and replace a leach without difficulty. The model indicates great ingenuity and effectiveness, but whether the saving of labor will overcome the disadvantages of cost must be determined hereafter.

The writer once tried a series of experiments on constructing leaches that should revolve, and while he had no difficulty in perfectly leaching the bark, he very much doubted whether the saving of labor was a sufficient compensation for the cost. The general plan consisted in making a leach eight feet square by two feet deep. This would, when constructed, make an inclosure which would hold one cord of ground bark, weighing about 5,500 pounds when wet, independent of the weight of the leach. This leach was hung by an iron gudgeon firmly screwed to the center of the flat sides or strong timber framework of the leach. On these gudgeons rested and turned this immense box or leach. Two of the ends were perforated full of fine holes, through which water or liquor was run or percolated for a limited time, and then by reversing the leach all the bark which had been packed would be loosened up, and the further percolation could be continued. The final pitching or casting of this leach was effected by opening a large trap door, which would allow all the bark to drop on the floor below, or it could be directed off into the stream by a shute. Six hours of active operation with this leach sufficed to take all the tannin from the bark. One of these small leaches would in twenty-four hours leach four cords of bark, with very little labor. But there is considerable machinery and wear and tear to this or any other kind of detached leach, and while they may be successful theoretically, practically the writer cannot commend their use, with his present experience.

CHAPTER VII.

HANDLING.

THE HAND REEL—THE ROCKER HANDLER—ITS CONSTRUCTION AND OPERATION—IMMERSED DRUM WHEELS—A METHOD OF RAISING HIDES FROM THE HANDLER VATS—THE TUB WHEEL HANDLER—HANGING HIDES IN THE HANDLERS—THE "ENGLAND" WHEEL—HANDLING WITH THE COX ROLLERS.

Passing by for the present the direct effect of the process of handling on the leather, in this chapter will be considered merely the *manipulation* of the stock, with reference to the amount of labor required, and the most approved methods for doing the best work at the least expense.

In no department of the tanners' art is the practice of different establishments more varied than in the matter of handling. Patents and improvements, almost without number, have from time to time been pressed upon the attention of tanners, and the methods of handling are to-day as varied almost as the tanneries are numerous.

Commencing with the old plan of "hauling up" by hand, throwing on packs, and allowing the sides to press and drain for a portion of each day—or the still more modern practice of "shifting" over from one vat to another—we have of late years adopted mechanical appliances which make this labor less a drudgery, and less exacting on the muscles of the arms, back and legs. The most generally approved methods are a "hand reel" and the "rocker handler." The hand reel is a revolving skeleton drum, which is made to turn on the top of a stand or frame elevated about three feet above the top of the vats. Both wheel and frame are light and portable, so that two men can easily remove them from one location to another, over the t ps of the different vats. The reel is placed on the alley-ways between the vats to be shifted, and the sides, hides or skins to be transferred, being tied together, are drawn over from one vat to the other by means of this revolving drum. One man is required to turn the crank and another to adjust the sides or hides in the head vat. By actual count it requires four minutes for two men to perfectly transfer 150 sides from one vat to another, and the labor is made as easy as the old method is fatiguing. Two men will average a pack every eight minutes during the day, including the transfer of the reel.

The "rocker" handler consists of a frame set in the top of the vats, constructed of wood. This frame should fill the vat within two inches on the ends, and one inch on the sides, so that when it rocks from the center it will play without touching. It should be made of two-inch plank and the frame should be of stuff 2 by 6 inches in substance. The end pieces should be of hard wood, not easily split, since they must bear the strain of the whole pack. The side pieces may be of pine or hemlock, but where hard wood is at hand it is best to use that for the whole frame. The center of the frame rests on pivots or steps, supported by uprights from the bottom of the vat. A "*stop*" at each end of the vat limits the rocker from vibrating up and down more than about eight inches. It is not intended to make any of these suggestions arbitrary as to construction or working, but a little thought at the beginning as to the most durable construction will save much repairing in the course of years.

The sides are attached alternately by the head and tail to

the cross pieces, backs up and bellies down, by means of hard wood pins, permanently fastened into the head frame pieces. Usually one end is fastened directly to or over the pin, and the other by an adjustable string, which may be not over a foot long, and may, by a permanent slipping noose, be used continuously.

[In the illustrations given on subsequent pages will be found good representations of both the "hand reel" and "rocker handler," from which their construction and the manner of operating them will be easily understood.]

This "rocker handler" will, also, in the next chapter, be commended over any or all other methods for making plump shoulders and offal, as well as for general economy.

Besides these two kinds of handlers now in general use (the latter in the lime and the former in the sweat leather tanneries), the tanners in and around Baltimore, Md., use a lattice drum fully immersed in the liquor; from the outside or circumference of this wheel or drum the sides are suspended, attached alternately by the head and tail. One method is to attach the leather to the outside, but the sides are generally suspended on the inside of the wheel or drum. In the first case the sides are reeled over the outside of the drum, falling in folds, much as the top of an umbrella comes together when put down, and those who use this method claim as an advantage that the liquor is pressed out in the process of turning and reeling up on the drum. But, as an objection to both of these plans, it may be said that they elongate the grain transversely, and make thin backs and thick offal. This effect is more noticeable if the leather is fully tanned in this manner, without being taken from the original vat in which it enters. If these immersed drum reels were used only as handlers, as is the case with the rocker, then perhaps the effects spoken of would not be so noticeable; but so observable are these defects now, that there is no difficulty in distinguishing leather tanned by this manipulation from all others, by the long grain transversely following the line of the back.

The liquors are pressed around to strengthen up the vats into which these revolving drums are used. The principle of the press in the case of these yard vats is not unlike the press of the leach already noticed in Chapter VI.; indeed, the same practice brings about the same result. Of course no agitation of the liquor is permitted while the press is in progress, otherwise the liquor would mix and destroy the principle of the action.

A most economical and efficient method of facilitating the raising and transfer of packs laid away in bark is effected in the following manner: A light frame made of plank is laid on the top of the liquor before commencing to lay away a pack. At the outset it floats on the liquor, but gradually sinks as the sides are laid on and "barked away;" when the sides are all in, the framework is at the bottom of the pack.

To each corner of this frame should have been permanently fastened a half-inch rope as long as the vat is deep. The loose ends of these ropes may float on an attached piece of wood, or may be fastened to the corner of the vat. When the pack is to be raised, the men, standing at each end of the vat with one of these ropes in each hand, together raise the whole pack to the surface, and when so raised they belay the ropes around iron bolts or wooden pegs, which are temporarily fastened at the corners of the vat. After they have reached down and removed say eight or ten sides from the top (all that they could readily reach without the aid of a hook), then they again raise the pack, and so continue until all the sides are thrown upon the pile, or shifted over to the next vat; but usually it is found convenient to pile the pack up on the adjoining vat, and lay away in the same vat from which the sides were taken.

The philosophy of this practice consists in taking advantage of the law of specific gravity. A side of leather will weigh less ounces under water than pounds out, and hence it is that two men can easily lift a pack while fully immersed in water or liquor, when they could hardly lift ten sides if surrounded by air. This system is more expeditious, is much easier for the men, and withal saves all opportunity for hook marks. Will harness and grain leather manufacturers please take notice of this latter advantage?

There is still another form of handler which, by way of designation, may be called the tub wheel or drum. It was first used at Sparrowbush, N. Y., in 1860, and consists of a large drum fully eight feet in diameter, and the width of the vat. This wheel revolves on a center shaft or flange and gudgeon, resting on the top of the vat. Nearly one half of this wheel is immersed under the liquor of the vat, and the remainder is, of course, out of the liquor. The stock is thrown in at a side trap door, and revolved inside this wheel until colored or ready to go into the layaways. Projecting wooden pins from the inside surface separate and carry around the stock. A more recent improvement on this wheel is to divide it off into four segments or sections, and thus make the compartments smaller. It is claimed for this improvement that the wheel turns easier, as a portion of the contents is always held near the center, and only goes to the circumference a portion at a time. Besides, a Swiss calfskin tanner told the writer that it was within his experience that the large wheel without partitions was too severe on green stock. He was probably right, for such violent action is apt to "purge" stock when it is green and the liquors are weak, as they always must be in the earlier process.

The hanging of sides, hides or skins over cross sticks held up by lateral supports in the vat, and the agitation of the liquor by *plunging*, is a most economical method of handling, and one that should be resorted to by all tanners who cannot conveniently adopt the rockers. The "strengthening up" may be effected by drawing off say one-quarter or one-third of the contents of the vat, and replacing by stronger liquor. The gentle agitation of the fiber and the exposure of the full surface of both grain and flesh to the action of liquor is essential to a proper handling process, and so long as this can be effected, no matter whether by manual or mechanical methods, it is all the same in effect.

The process which has for so long a time been improperly attributed to Mr. England, and known as the "England wheel" method, has very general acceptance by many light leather tanners. In this method of handling the wheel revolves on the top of the vat, and is constructed in skeleton form, with buckets or paddles, which, when turned, dip about six inches in the liquor, causing a thorough and constant agitation of the contents of the vat. If the vat is square in the bottom there will be a great agitation, and it will answer imperfectly the purpose, but to make the liquor and the stock contained in the vat revolve at the same time with the greatest facility, the bottom of the vat should be circular, but *not latticed*, as was contemplated by the patent of Mr. England.

This description of a circular-bottomed vat with a revolving wheel in the top had its origin in the city of Brooklyn about the year 1840; but circulating or handling vats, in a modified form, were well known in Germany long before they were introduced here. For the coloring of sheep, calf, kip, light upper or splits, this wheel and circular-bottomed vat is most serviceable, but when applied to the after process of tanning, as has been attempted, especially in heavy leathers, must always result in failure, according to the experience of the writer.

Mechanical power should always be introduced into the handling yard. All the mechanical appliances which have been mentioned are capable of being operated by shafting, pulleys and belting. The "rockers" and "skeleton drums," particularly, should be driven by power, which may be accomplished by allowing a shaft to run overhead and between each tier of vats. An "eccentric" or "crank" motion can easily be geared from a shaft, as all mechanics understand.

The writer has not called attention to several attempts to raise whole packs, both in the handlers and also in the layaways, by mechanical power, because they are generally expensive and ineffectual methods of handling.

Perhaps this list of handlers would not be complete without including the patented rollers of a Mr. Cox, of London, which were brought to this country about the year 1840, and were adopted only in one or, at most, two yards, but, as they were in quite extensive use in some portions of Great Britain, they deserve this mention. The rollers are made of wood, and are covered with cloth or matting; two rollers, about two feet in diameter each, are made to revolve with their surfaces so close together as to draw the side through and squeeze out the spent liquor, which, according to the theory of this patentee, has deposited its tannin, and should be displaced to give opportunity for new and freshly charged ooze to come in contact with the gelatine of the hide. The sides or butts are tied together in such way as to make a continuous belt, and by means of these compressing rollers are drawn from one vat and dropped over into the next.

The effect of this frequent compressing of the fiber is to make the leather thin and hard. The writer has seen some very heavy weighing leather made by these rollers, but as the system contemplated and used much stronger liquor than it was usual for our tanners to employ at that time, to this cause was attributed the wonderful resulting gain. In the first experiment with these rollers at one of the tanneries in this country 194 pounds of leather were made from 100 pounds of dry Buenos Ayres hides, but the rollers were so expensive that they have long since gone into disuse here, and measurably so in England; yet it must be confessed that for butts, bends and trimmed leather, it is much better suited than to our sides, and to supply a demand for firm and even hard leather, as is required in Great Britain, these rollers may serve a useful purpose.

So long ago as the year 1842 the writer saw at Woburn, Mass., a process of handling upper leather on frames, immersed in the vats, which at that early day presented a foreshadowing of the more modern and better process of rocking the handlers. But even yet the stationary sticks or frames are in use by pebble and grain leather tanners in several large tanneries. This successful practice only demonstrates what the writer has repeatedly had occasion to say, viz., any method by which the fiber is gently agitated and the whole surface exposed to the liquors will answer the purpose contemplated.

CHAPTER VIII.

HANDLING AND PLUMPING.

THE USE OF VEGETABLE AND MINERAL ACIDS—THE EARLY USE OF VITRIOL BY AMERICAN TANNERS—CONSIDERATIONS AFFECTING THE AMOUNT WHICH MAY BE USED—ITS EFFECT ON LIMED AND SWEAT STOCK—STRENGTH AND AGE OF LIQUORS TO BE USED IN THE HANDLERS—DIFFERENCES IN THE HANDLING AND PLUMPING OF SOLE AND UPPER LEATHER.

The treatment of sole leather is so different from that of upper, harness, etc., in the preliminary or handling process, that the work on these two classes of stock will be considered separately.

It has become of late years quite universal to make a distinction between "acid" and "non-acid" treatment in preparing hides to be tanned into sole leather. These terms, although in common use, do not convey an accurate idea of the differences in the process. No sole leather is prepared for the yard without the use of acid of some kind, but what is meant by the term "non-acid" is, that only gallic acid coming from bark liquor is used in the one case, and in the other sulphuric acid or vitriol is employed. This employment of vitriol as a means of plumping the fiber of the hide is of quite recent date in this country. In Great Britain the distinction has been maintained for a long time between "vitriol" butts and bends and those that are not raised by this process. As early as 1853 the writer commenced the use of vitriol in the preparation of "union crop" leather. But this was on limed stock, and in very moderate quantities. The mineral acid was used quite as much to counteract and neutralize the lime as to plump the fiber. When used to this limited extent the "buff" was not injured, the grain was plumped, and the whole structure of the hide slightly swelled. But from this small and cautious beginning the use of vitriol extended until the sweat leather tanners took it up, and from about 1860 to the present time it has come into general use.

If this practice had been intelligently and openly introduced, far less damage would have resulted; but each tanner, depending upon his own judgment, and, it must be confessed, on very limited information, has gone on blundering, all the time thinking himself much wiser and more skillful than his neighbor, until vitriol-raised leather has come to receive the condemnation of all of our best manufacturers. Certainly it is discredited by all who have regard to the appearance of the bottoms of their boots and shoes.

For several years after the first introduction of the new practice the fact of the use of vitriol was kept a secret, and was not exposed until the general quality of our sweat leather had deteriorated to such an extent as to cause alarm. Now the proper use of this acid is better understood, and the early abuses in its application are measurably avoided; but even to this day many otherwise good tannages are injured by the improper use of vitriol in the handlers. In this, as in all other innovations, there should be more frankness among the tanning fraternity; its members should learn that ignorance and bad workmanship reflect finally as much on the innocent as on those who are more directly responsible. A tanner who makes poor leather, particularly if the defects are latent and not at once discernible, unfavorably influences the whole trade.

The writer would gladly impart such information in regard to the proper and permissible use of vitriol as would help tanners to avoid the defects complained of, but the circumstances under which this acid is employed are so varied that any general formula or direction for its use would be almost certain to mislead some of those who might wish to be guided thereby. Hides are not all of the same specific nature; some will bear more acid than others. Thick hides will bear more than thin ones; steers more than cows; the butts more than the bellies of the same hide. The consequence is that when our American sweat leather tanners use vitriol on hides worked in without proper classification, and without trimming, they are apt to overdo the thin sides or thin portions of the thick ones. There is no way to avoid this difficulty without taking care only to strike the thin portions fairly through and then stop, leaving the butt only partially impregnated.

In regard to the experience and practice of our sweat leather tanners in the use of vitriol—some color their sides before they introduce them to the acid; others use the acid first and color last, while still others, and probably the majority, use both agents at the same time—they color while they plump—*i. e.*, they use the acid in the coloring handlers, and strengthen up the handler as each new pack goes in, with new, sweet, highly colored liquor and vitriol.

How much vitriol may be used with each pack? This question is often asked, but can never be definitely answered without a knowledge of the hides, and of other circumstances. It must depend upon how much the previous pack has taken up; how strong the acid is; how heavy the sides are; how fully the leather is allowed to carry off the acid, etc. There will be no difficulty experienced if tanners will regard vitriol plumping as a tawing process. If they will cut the thin portions of the sides, and examine in a strong light, they will notice just the extent to which the acid has penetrated; the progress is as clearly defined, although not quite so plainly marked, as if the tan liquor itself had been present.

Vitriol preserves the hide, and holds, when combined with salt, the gelatine from decay. All the green sheepskin pelts that come from England to this country in casks are preserved by this process, and may be so held without damage for years. When these skins come, however, to be tanned, the acid must be neutralized before they go into the liquor, or while in the liquor, otherwise a result is produced just such as we see in much of our acid sole leather. A sheepskin tanned without neutralizing the acid will swell to more than double its wonted thickness, and will tear as easily as brown paper of the same substance, while the whole structure assumes a dark brown hue, which grows darker as it is exposed to the air. From this fact tanners may take a hint, as the writer has done. Indeed, from some early experience in tanning what are termed "salted" sheepskins, much of my early knowledge of the use of vitriol was obtained.

Vitriol distends and injures the fiber, and if not removed from the grain will deposit immediately underneath it a dark brown strata, which is most objectionable to manufacturers who use buffing wheels instead of hand labor in removing the grain on the soles of boots and shoes.

In my judgment, vitriol should *never* be used to raise purely sweat stock. For hides prepared with lime we all know that vitriol to a limited extent may be used without material damage. What, then, is to be done? All sides as they come from the sweats should be thrown into a lime, and there handled until they fairly begin to feel its effects, say, ordinarily, about six or eight hours. The lime will, in this time, sufficiently penetrate the surfaces to neutralize the acid when the sides reach the handlers. Thus treated the dark strata will be either entirely removed, or so far modified as to be less objectionable, depending on the strength of the lime and vitriol used and their relative proportions.

It should be remembered that sole leather raised with gallic acid (if the acid be strong) will leave a modified dark strata, but this is controllable by other acids, which the boot and shoe manufacturers employ in "cleaning up their bottoms." No counteracting agent has yet been found, however, for the vitriol stain, and the exposure to the atmosphere of leather thus affected increases its unsightly appearance.

The German-tanned "warm-sweat" sole leather has the difficulty here complained of to a more marked and objectionable extent even than our own tannages; but as the consumers there care less for the buff than we do, this defect seems to be disregarded. They use vitriol or acid liquors even to a greater degree than we do, and to such an extent indeed that most of their sweat sole leather will crack and even break by close folding. Consumers of leather in Continental Europe, having been accustomed to this defect in their own home supply, make less objection to it in us than do the manufacturers of Great Britain. The tanners of Great Britain do not sweat, either by "cold" or "warm" process, and hence they use vitriol and lime jointly, with less damaging effect.

It is proper here to say that in France, Germany and Switzerland women's sole leather and all the light leathers are limed, as in Great Britain, and this "cracking" and "old hatty" appearance of the fiber is confined to the sweat vitriol and acid raised sole leather, which are greatly inferior to their other tannages.

Vitriol-raised leather, when treated in the after process with weak liquors, produces a most unsatisfactory result.

The grain is poor, the fiber coarse and hatty. No attempt is here made to solve the mooted question whether vitriol does not destroy the tannin. It is conceded that it will plump and hold the fiber, and will facilitate the tanning, but whether these advantages, in an economical point of view, are not more than overbalanced by its distinctive action on both fiber and the tannin, must be left for future experience to determine.

In closing the suggestions on this point the writer would like to say that great care should be exercised in handling while in the vitriol vat. If the acid is allowed to settle down on the grain, stains will result which will cause a *clouded* appearance that will remain to the end of the tanning process. My own judgment is that it will always be serviceable to hang the sides on cross sticks and plunge as the most certain method of distributing the effects of the acid uniformly. In adding vitriol to the pack, be sure to put it in before the sides go in, and plunge thoroughly, otherwise the acid will drop down and discolor the grain. The specific gravity of vitriol being much heavier than that of tan liquor, its tendency is to settle, and unless great care is taken it will so settle and mottle the grain of the pack.

In preparing hides for sole leather by the lining process, if only so much lime is allowed to permeate the fiber as to enable the tanner to remove the hair, as has been already indicated, distinct bating is unnecessary. If the hide is filled and swelled with lime, the treatment should be such as to remove such excess of lime before the sides come to the handler. But, usually, if the sides are wheeled in warm water for ten or fifteen minutes, they will be sufficiently cleansed to enter the handler.

The first handler liquor should be the oldest and most worthless liquor in the yard, so worthless that after handling the green pack in it for a few hours it can be run off in the stream or sent again to the back leach.*

The pack, after being thus cleansed, so far as this old spent liquor will do so, takes its place in the order of succession in the handlers. If, as is assumed, the rocker system of handling is observed, then it will take the place of the pack which has just gone to the layaways, and a liquor is prepared specially by pumping the strong liquor found in the vat over on the next pack, and so on around until the weakest liquor is placed on the new incoming pack; care should be observed to feed each pack with an increasingly strong liquor, not, as too many practice, once in each day, but at least twice.

Far more damage is done by feeding too slow than too fast; so long as the old liquors from the layaways are used there need be no fear of either feeding the packs too fast or too frequently, only let them be fed in their order of entrance.

Of course, as many packs should go out as come into the handlers daily. It will be found profitable to keep each pack from ten to twelve days in the handling rockers, or in any other method of handling that may be adopted. This time will suffice to overcome all the effects of the lime, fairly plump the sides, and set them well on their way toward the after tanning process.

As a last parting word on this subject the writer would urge tanners to put away from their minds the fear so often expressed that liquors may be too strong to start with. If eight or ten packs are in the handlers, commencing with the head

* Just here the writer would say that, whether this old liquor should go again to the back leach or off in the stream will depend very much on the manner of using the liquors. If, as in the true system of using the press leach, no liquor is allowed to go back until expended, then the dirty water (for it is little more), without acid or tannin, which washes and cleanses each new pack which comes to the handlers, should be discarded; but if there should remain either gallic acid or tannin in the liquor, then it should certainly go to the back leach. Without exercising some discretion in this regard valuable qualities may be thrown away.

pack and running down to the last one, the best and strongest of the discarded layaway liquors may be used with safety. These liquors will always contain acid enough to prevent the "puckering" or "binding" of the grain. Liquors weighing by the barkometer 16° are none too strong for the head handler. So large a proportion of this weight will be acid that they may safely be trusted to work no harm.

The treatment of the lighter tannages, such as upper and harness leather, in the handlers, is far different from that of the heavy sole leather tannage, just considered. These leathers are sold by the side or foot, or should be, as no tanner can afford to treat his stock as such goods should be treated, and then sell it by weight.

After the lime has been thoroughly removed by a bate of hen manure, or wheat bran, or some other depleting process, the stock enters the handler, in the first instance just as the sole leather did, but instead of taking a course of old, sour, strong liquors, it is only retained long enough in the acid liquors to "brighten" and clear the grain, and then weak, sweet liquors should be fed to the packs, just in the order in which they come from the leaches. These liquors should go back to the leaches very frequently, to be "sweetened" and "strengthened" up. Fresh sweet liquors standing from 10° to 12° should be as strong as is ever permitted in the handlers. The "rockers" should be used, however, to plump and handle even upper, harness and calf, for, although the sweet liquor will prevent that extreme plumpness so desirable in the heavier tannages, a close, round grain will be formed, and a degree of fineness and plumpness of the shoulders and offal will be secured not attainable by the usual manual handling process.

CHAPTER IX.

LAYING AWAY.

TIME REQUIRED AND STRENGTH OF LIQUOR WHICH SHOULD BE EMPLOYED—TANNING IN THE HANDLERS VS. LAYING AWAY—EUROPEAN METHODS—"BLACK ROT" AND WHITE SPOTS—THEIR CAUSES AND THE REMEDIES—SHOULD HIDES BE LAID AWAY GRAIN UP OR FLESH UP?—MAKING WEIGHT IN THE LAST LAYER.

Proceeding in the order of the successive processes of tanning, we find our pack coming from the handlers not only well "plumped" and uniformly "colored," but fairly "leathered." The stock has taken upon itself "substance," and a reasonable degree of firmness. Supposing that four layers will now carry out our tanning and completely fill the fiber, we shall assume that the remaining work will be divided into the following periods of time, during which four successive layers of bark will be given each pack. To the first layer we will give ten days, to the second fifteen days, to the third twenty days, and to the fourth thirty days, making in all seventy-five days. But it must be evident that leather differing in substance and weight will considerably vary in its requirement of time; and then, too, much will depend upon the season of the year, the strength of liquors applied, and other circumstances. If we commence with a liquor of 16° degrees strength with the first layer, and end with 30° on the fourth layer, seventy-five days will be found ample time to tan the heaviest sole leather.

The writer tried an experiment on one pack of slaughter

hides, to test the actual time necessary to tan good middle and overweight leather, averaging say 20 pounds per side. This experiment, made under his own daily observation, enables him to speak with great confidence of the result. The hides were bought of the Butchers' Association of the City of New York, and weighed 70 pounds each, trimmed and cured. One week was consumed in their preparation for the bark. They were colored and handled for three days in an old sour liquor from the first layers, when they were put into a liquor of 20 degrees strength and handled daily, "shifting" from one vat to another each day, and receiving just enough added strength to keep the liquor at 20 degrees. At no time was the liquor over 22 degrees or under 18 degrees. In sixty-one days from the time the sides came into the handlers they were taken out not only fairly "struck," but well tanned. The color was "fairly good," although not as bright as other packs which had been laid away in the usual manner, and had taken fifteen to twenty days more time. The conclusions drawn from that experiment were as follows:

1. Well cured New York City hides will make 65 pounds of leather for every 100 pounds of green cured and trimmed hides.

2. Sixty days in the actual process of tanning, handling once each day in suitable liquors, will tan the ordinary middle and overweights sole leather.

3. About one-third more time is required if the leather is tanned by being laid away in the usual manner.

4. The color is both lighter and more uniform when laid away than when handled. Besides, the tendency of the sides to bag is far less when tanned in the usual way, by laying away in bark.

This experiment, if read by a German, Swiss, French or Austrian tanner, will be regarded as extraordinary, but an account of it is given here, not because it is exceptional in this country—for many better results have been attained—but for the purpose of bringing into contrast our methods of tanning as compared with those of the Continent of Europe, for in all statements affecting results the writer must be understood as drawing a clear distinction between the processes in Great Britain and those of the Continent.

It is not claimed that these results are practically secured in all instances, or even in a majority of cases in this country, for the causes of detention which go to make the average time and results are but too well known. These causes of delay will be considered in their appropriate place, but theoretically our system accomplishes all that is here claimed.

As possibly these statements may be read by tanners over the sea, for their information, as well as for the instruction of our own tanners, it may be said that our universal custom of leaching new bark and using the decoction obtained for our tanning agent, is not practiced to any extent by tanners in Continental Europe. They color their green packs in handlers made from the old sour liquors, which are pumped from the layaways, adding (as is the custom of some) a few baskets of "spruce" bark. The hides, in small yards that are run without power, are "poled" in the vat every hour, and once or twice each day the stock is handled up. But while yet quite "green" and "pulpy" the packs are laid away in outdoor round tubs or vats, from eight to twelve feet deep, and from six to ten feet in diameter.

The first layer is made of old, partially-spent tan, laid fully one inch thick on either side of a hide or skin, and well stamped down. When the vat is finally filled with these alternate layers of bark and green pelts, water is run in to fill the interstices. The acid of the bark preserves the gelatine, and at the same time neutralizes any lime which may have been left from the handlers. After twenty or thirty days' time this stock is raised, the bark "skimmed out," and the pack relaid. To the old a portion of new bark is added, and at each subsequent "shift" or "turn" of the pack more new bark is added, all the time holding the old acid liquor which has accumulated.

The writer was informed that in most instances from sixteen to twenty months are occupied in these layers before the leather is finally declared tanned, and the statement may be fully credited, for in no instance, among the numerous tanneries visited, did he see any attempt to leach new bark; such leaching as was attempted was confined to the old, spent bark, that had been literally worn out by frequent handling. This process is slightly varied with calf and upper in the handlers and layaways, but substantially the same system is followed out in all their tanning. Perhaps, in justice to a few large calfskin tanners—notably so in the case of M. Mercier, at Lausanne, Switzerland—it should be said that much more vigor and a greatly improved system was employed. This will be more fully mentioned in the chapters on calfskin tannages, in another place.

The writer hesitates to state the methods employed in Great Britain, for the tanning agents there used are so entirely different from our own that comparisons are impossible, and might, if given, be misunderstood. They use decoctions quite as strong, in fact, as we do, and much stronger, measured by the test of the barkometer, while the time in which they tan is very nearly or quite equal to that made by our best tanners. If they take longer (as many of them do), as much even as ten months, it is on heavy South American hides, made into butts and bends that are very stout in substance and extremely fine in texture. But native sole leather hides, also upper and kip leather, from both native and foreign hides, are "put through" quite within our best time, and in the most artistic way. American tanners can learn much from the economies practiced and results produced by the tanners of Great Britain, for they do manage to make exceedingly good leather from very small quantities of tanning material, whether native or foreign.

To return from this necessary digression our methods of laying away will now be given with more minuteness. More than forty years has passed since the "kiffing" away process was given up, and leaching and bark liquors introduced in its stead. Few men are old enough to remember when at Salem, Danvers and Cummington, Mass., water was used instead of liquor in the layaways for sole leather tanning. At present, the bark liquor used in the layaways is depended upon to do the tanning, and not the interlying bark at all.

The general usage now is to run the strongest and newest liquor direct from the leaches upon the head packs of the last layer. These liquors should weigh by the barkometer fully 30 degrees of strength. After 30 days' use they will be reduced to 24 degrees. A portion of this indicated strength is acid. Twenty days' further use on the third layer will reduce the strength to 18 degrees, and thus down step by step until a large portion of the strength will be acid and not tannin, when the liquors will be put on the first layaways, or may be run into the handlers and fully exhausted.

This is not, it is true, an invariable custom, for too few of our tanners have anything that may be called uniform methods, but the general leaching of bark and the use of the graded decoction on the packs, in the order of their coming in and going out of the yard, must be conceded to be our common practice, and to this extent the system is exclusively American and English.

The manner of running liquors from the leaches to the yard

is by covered, inclosed wooden tubes or bored-out logs, either under ground or under covered ways. The method of construction and cost of these will be stated hereafter, but where heat is used on the head leach, these liquors are too often sent into the yard warm, or even hot, and when thus used, very much injury to both color and buff is the result. Yet, as these head or strong liquors are, or should be, confined to the head or nearly tanned packs, positive damage to the gelatine of the hide does not follow to the extent which would be supposed; while if these warm liquors, from mistake or ignorance, go to the half-tanned or green packs, "black rot" is almost sure to result, particularly in the summer season. The damage is caused by the decay of the animal fiber—actual decomposition of the untanned or raw hide. The external appearance of this damage appears as a sinking in of the two surfaces, caused by a general disturbance of the central tissues. When the decomposition proceeds still further, and the surfaces break, partially tanned "puss" exudes. This disturbance and damage has been popularly denominated the "black rot," because, upon the grain side, in addition to the "falling away" appearance spoken of, the surface turns black, showing dark or black spots wherever this damage occurs, and this is generally in the thickest and best portion of the hide.

While on this subject of latent defects and damage, perhaps it is as well to state the experience of our tanners in regard to the "white spots," of which there has been so much reason to complain, and which, even with our conceded ability to remedy, still cause great depreciation of value in much of our sole leather. This defect occurs from imperfect beam work. In the sweating process the hides are soaked, milled and sweated within a week, and when the atmosphere of the sweats and the nature of the hides favor, the hair "comes" quite freely; as that fact, under our process, indicates that all has been done that is necessary, the stock is turned into the liquor, when, in fact, the workman has never touched much of the grain surface of the hide, thus leaving mucus and grease. Often the grain has been saturated with the greasy substances of the animal, which are not overcome or worked off by the beam hand, but remain to resist the action of the liquor. Some tanners call this "mucus," but the greasy substance which causes this white spot is different from mucus, as will be demonstrated by a little examination.

The New York leather trade, several years ago, offered a reward to any tanner who would discover a remedy for this defect, and a Mr. Edson came forward with a solution of the difficulty. He discovered that potash, soda ash, or any strong alkali, applied to this untanned or uncolored white spot, would at once make the grain take the coloring matter, and subsequently receive the tannin. This was a perfect remedy, and led, as any one may perceive, to the cause of the trouble. Whereas other tanners ascertained that rubbing the surface with pumice stone or brick, or scraping thoroughly with a knife, would measurably relieve the damage, the real cause of the difficulty was not suggested until the alkali test was applied, and then we all comprehended that *grease*, in some form, was the source of the trouble, and, therefore, to Mr. Edson was awarded the valuable consideration offered by the trade. If, therefore, hereafter, any tanner suffers from the appearance of these white spots, and does not apply the remedy, he has only himself to blame. These spots may be avoided in two ways: First, by thorough working on the beam; or, second, by the use of a strong alkali on the spots affected, after the sides have been colored, and the defect complained of discovered in the handler.

The best practical method to overcome these spots would,

perhaps, be for the tanner to have a carboy standing in his yard with either a strong decoction of sal soda or soda ash, from which the attendant on the handlers can take a small portion, and by the use of a rag or sponge touch, and even slightly rub, the affected part. It will cause a dark or even black stain at first, but this will finally disappear in the after process. Some tanners prefer to throw the sides affected on a table or beam, and rub or scrape the spots with a smooth steel edge; a round pointed knife is serviceable for this pur-purpose. The latter course can only be pursued, however, when the spots are small and infrequent. My judgment is that where sufficient care is taken to work fully over the grain in the beam house these spots will not be encountered in the handlers.

It is probable that if hard wood ashes, soda ash, or sal soda, were freely used in the soaks of such hides as usually give this trouble (as, for instance, dry Western and California), no white spots would ever make their appearance in the handlers; besides, the greasy hides would soak more uniformly. When a pack of dry hides is thrown into soak indiscriminately great injustice is done the "sun dried" and greasy portions, from their more persistent resistance of the soak.

The question has often been raised whether the sides should be laid grain or flesh up in the layaways. The practice is to lay grain up, and this is justified on the ground that, in "hooking up," the grain is not so likely to be scratched and marked as it would be if the sides were laid flesh up. If, as seems quite likely, the color is seriously affected (particularly in hemlock tannages) by the settling of the coloring matter on the grain, and a deeper, darker red is the result, then tanners may well inquire whether, in their attempt to avoid hook marks, they do not entail upon their stock a worse evil. Besides, it may be asked in this connection whether it is not probable that tannin will enter the fiber of the hide more naturally from the flesh than from the grain surface. The pores of the hide, when on the animal, certainly do open their valves outward from the flesh, for all the emanations of the animal body go through these outward opening valves or pores of the skin, which never receive back from the grain to the flesh. We all know that the goat and sheep skin tanners, when they "sew up" their skins, keep the grain outward, and tan altogether by pressure from within. But some one may ask, Does not the tan liquor surround both surfaces, and seek admittance equally from each side? This may be answered in the negative, for the heavier or stronger liquor settles, and when the sides lie upon each other in layers, both the strength and the coloring matter tend to settle downward. The theory is that when the tannin comes into contact with the gelatine of the hide the union is made instantly, and then, if both the hide and tannin are allowed to stand in perfect quiescence, the conclusion is drawn that not only does the strong liquor settle, but it does its work instantaneously, from underneath, while from the top of the side this specific-gravity principle of the liquor keeps on acting for a considerable time, forcing downward the renewed strong liquor, and bringing itself into contact with the untanned fiber. This may be refining on speculative ideas, but such a theory does prevail among some.

One of these theorizing tanners, within the knowledge of the writer, always acted upon this idea, and made it his habit, several times each day, to walk over his layaway packs, stopping to press his weight forcibly on each as he passed. He always insisted that by this course he displaced and disturbed the fixed relation of the liquor to the gelatine of the hide, and brought new tannin in contact with the hide. This we know, that agitation of the fiber and frequent replacement of tannin does facilitate the process.

The last or fourth layer is purposely prolonged, not only to fill the fiber and make the leather firm, but also to brighten the color, as is sure to result from the acid liquor, which accumulates with age. The fiber is fairly tanned on the third layer, but the filling process—the extra weight, indeed—is made on this layer, and the tanner who fails to give ample time to this last layer must be content with inferior solidity and gains. Can leather be overtanned? Yes. It may be so thoroughly tanned as to leave no grain to buff, and really no life or elasticity to the fiber.

CHAPTER X.

DRYING AND FINISHING.

WASHING AND SCRUBBING THE LEATHER—THE "HOWARD SCRUBBER"—WHEEL OR DRUM SCRUBBING—DRAINING—HOW THE ADMISSION OF LIGHT AND AIR SHOULD BE REGULATED IN DRYING—DAMPENING BEFORE ROLLING—THE FIRST AND SECOND ROLLING—EFFECT OF THE ROLLING ON THE BUFFING QUALITIES—BLEACHING WITH SUGAR OF LEAD AND SULPHURIC ACID—THE WARM SUMAC BATH—EFFECT OF THE LATTER ON CALFSKINS, GRAIN LEATHER, ETC.

From the last layer in the yard the finishing process begins. If mistakes have been made and defects are apparent it becomes the duty of the finisher to consider and overcome them. Where a uniform system prevails, both in the beam house and yard, such defects should not be of frequent occurrence; but whatever they are, and wherever they occur, it is the duty of those who have charge of the drying and finishing loft to make amends, as far as possible.

The first duty of the finisher is to cleanse the leather from all sediment and extraneous matter. This can be better begun before the leather has been exposed to the air than after such exposure. For this reason the pack should be taken in hand immediately on coming from the layer, when the leather should be thrown into clean water, or may, if there are stains and a mottled condition of the grain, be thrown into an old sour liquor, and left for a day or more. We must bear in mind the object to be attained, namely, to cleanse and purify the grain and flesh. This can be done in various ways—first, by rinsing and hand scrubbing; second, by revolving drum wheels, with surfaces covered with splint brooms. This method is covered by the patent of Mr. Howard, and the scrubbers are known as the "Howard scrubbers." The third plan is by a revolving drum wheel, during the revolutions of which the leather is washed by a constant tumbling and turning over in water, which is freely supplied. This wheel, in its application to such service, is also covered by a patent. The merits of each of these processes will be considered in their order, but before presenting the remedy for the evils which necessitate this scouring let me consider their cause.

If the liquors are obtained from the press leach there will be no sediment or discoloration to remove, and consequently nothing more will be required than rinsing and scrubbing with a hand broom, or brush. But if the liquors have come from the "sprinkler leach," or from the ordinary "flooding leach," then there will have accumulated bark dust and sediment, blended with resinous matter, which will attach itself to the flesh and grain, and will require mechanical power for its removal.

The "Howard scrubber" is now largely relied upon to remove these defects. It consists of two revolving rollers or drums, with their surfaces coming very near together, to which are attached brooms or brushes, which quite meet. Through these scrubbing surfaces the sides are made to pass—sometimes once, but often twice and three times. The friction is increased by holding back the side with the hand with more or less firmness. Two men can, with this machine, pass through about 500 sides per day, and do the work well.

More recently the "wheel" or "drum scrubber" has come into use. This consists of a drum wheel about six to eight feet in diameter, four or five feet wide, made strong enough to resist the hammering process of the revolving sides of tanned leather, which are made to "thrash" around in the inner surface of the wheel as it revolves slowly, say about twenty revolutions per minute. The inner surface of this wheel is furnished with pins or projections, which carry the sides around with the motion until they are raised sufficiently from the bottom to violently agitate the fiber by their dropping down, and by the friction caused by the sides rubbing against each other the surfaces are cleansed, very much on the same principle as iron castings are now made smooth by being rubbed against each other in a revolving wheel—or as shoemakers' pegs are polished by friction and contact—being revolved in bulk inside a tank or hogshead. It is claimed that the most of the bark sediment or accumulation is removed by this wheel within five minutes, and as about ten or fifteen sides can be thrown in at a time, it will be perceived that the cleansing process is rapid and economical.

Great care, it appears to me, should be observed lest this wheeling be continued too long, the effect of which would be to pound out the weight and make the leather soft. The writer is assured that with care these defects can be avoided.

When the sides have been thoroughly cleansed, by means of either of the foregoing processes, they are usually laid in packs to drain, but in doing this great care should be taken to place flesh to flesh and grain to grain, with each side so exactly covering the other as not to allow the flesh of one to touch the grain of another. Unless this precaution is taken the strong liquor, which is absorbed and held by the flesh in undue proportion, in spite of all attempts to remove it, will impart itself to the grain of the side it comes in contact with, if the pack is unevenly spread, so that the drainings can run down and stand in small pools on the grain surface. Such spots will be marked with the stain of the bark liquor when the leather is finished.

The most ready way to avoid the forming of these "cups" is to throw the pack over a "half log," cut lengthwise, which is made as follows: Saw a hemlock log, two feet in diameter, through the center; turn the concave surface up and the flat surface on the floor, and lay the pack lengthwise on this stick of timber; it will be almost impossible in this way, with ordinary care, to have any cavities form into which the spent liquor will run and cause these stains.

The pack, while thus exposed to the air, should be covered with canvas, to prevent the edges and exposed surfaces from meeting the light and air; otherwise such portions will become darkened to such an extent as to show finally great discoloration.

After the pack has been thoroughly drained, a slight coat of fish oil should be put on the grain and flesh by a rag, and the leather may then be hung up to dry. If the Turret dryer is used, or the principle of that dryer is observed in the construction of the building in which the leather is placed, the perfect control of the light and air thus obtained will enable the workman to keep all light and any considerable drafts of air from the leather for three or four days, or until the sides are fairly stiff and dry on the surfaces. After the moisture has been evaporated from the surfaces air may be admitted, but not a strong light, for the free admission of light to the leather is certain to darken the color. Ten days will dry the heaviest sole leather if a proper drying loft is employed, but, whatever time is requisite, the leather in all its parts should be thoroughly dry before it is taken down. When in this condition it may be held subject to the call of the roller.

Two days at least before the leather is rolled, the "dampening process" should begin. The sides should be first sprinkled with water carefully, both on grain and flesh, and then laid down, grain to grain, and flesh to flesh, taking care as before not to have "cups" form where liquor stains may occur. After a period of a few hours the same process may be repeated, but with more care, making certain this time that all parts have not only been reached, but that no more water has been applied than will readily absorb. The pack should then be carefully laid in a large wooden box or tight room, care being taken to pack snugly, so as to get in as many sides as possible, and at the same time to prevent the air reaching the skirts to dry and discolor them.

If the dampening process has been attended to with care, the leather will be in condition to "pack" under the roller without "rebounding" or "coming back;" but to make sure that all the parts are evenly moistened, great care should be taken by the attendant to re-sponge both grain and flesh lightly before the sides are passed to the roller, taking special pains to retouch such spots as have lost their moisture, and become partially dried. It will also serve a useful purpose to pass an oiled rag or sponge over the grain surface, to prevent "furring up" under the roller and on the roller bed. The leather should be so prepared that the whole fiber will pack solid, without being so damp as to induce the soluble portion of the coloring matter to press through the grain, as will be the case if the leather is rolled when too damp. We all know that snow may be too dry as well as too damp to pack, and the same is true of leather. The happy medium should always be observed to secure the best result.

Since our tanners have been trying to meet the tastes and wants of the German market, some of them have adopted a somewhat different method in finishing vitriol raised leather. Only the grain is dampened, and the residue of the fiber is rolled dry, or nearly so. This process gives a fictitious substance, and a harsh, hard, dry fiber, which seems to correspond to the character of the German tannages, but cannot permanently meet the wants of either American or English manufacturers.

All sole leather should have the stretch taken out of it so completely that when the sole is cut by machinery there will be no waste, as it goes on the bottom. It should exactly fit, without any paring or loss.

Leather may be too hard as well as too soft, but it cannot be too solid. Keeping in view the distinction between "hard" and "solid," the reader will understand that solid leather is a well-packed fiber, which cuts "cheesy" and "smooth," and not dry and "husky," as much of the vitriol raised leather cuts, increasing in these undesirable qualities with age.

The leather is rolled for the first time in the moist ("sammied") condition previously described. The grain is fairly flattened and made smooth; the whole fiber is firmly packed; but there remain roller marks on the grain, and defects which must be overcome. How shall this be done? Some will say —let the leather be fully dried and then brought back and rolled on the flesh side, grain side to the roller bed, and this will remove all marks on the grain from the first rolling, and will, besides, leave a gloss and finish on the grain which is most desirable. Much can be said in favor of this style of finish, and, perhaps, no one has been more influential in introducing it than the writer. But for all that, it is not the most artistic or desirable. It will do on hemlock leather, where mere color in the buff is sought, but where both "color" and a "velvety" buff is desirable, *i. e.*, where the bottom, after being buffed, is to present a "soft nap," and at the same time a lively and beautiful flesh color, then some other process must be adopted. This other and better process is a second rolling, immediately or very soon after the first. The sides, after the first rolling, may be spread about the loft, and within an hour after the first rolling the second rolling should take place from the flesh side, as the first was from the grain. It may even be of service to slightly redampen the grain in spots, if the leather is allowed to remain as much as an hour in a dry atmosphere between the first and second rolling.

The care required to finish "union crop" leather, as well as pure oak, is incomparably greater than is usually bestowed on the ordinary hemlock, and yet the time is probably not far distant when the same care will fully compensate the tanner in the latter as in the former tannage, particularly where slaughter hides are used. Some of our union leather tanners are now turning their attention to pure hemlock tannage, and their greater success in producing hemlock leather justifies the impression that more care in the finish of slaughter hemlock will well repay the tannner. One continent has come to be well nigh convinced that hemlock bark is as serviceable in tanning as oak, and it is quite within the range of possibilities that another decade of years will bring the whole Eastern World to this conviction.

The branch of the finishing business which may be called the bleaching process might profitably occupy a chapter by itself, but as the writer does not pretend to fully understand —and, if he did, would not commend the various devices of bleaching leather by sugar of lead and sulphuric acid, which is the most common practice in general use among fair leather manufacturers—the suggestions on this subject will be compressed in this chapter. The practice now is to dip the sides alternately, first into a bath of sugar of lead, and then into one of sulphuric acid, until the coloring matter of the hemlock is fully removed. This bleaching process produces an immediate effect that is almost magical, but when the finished leather is exposed to the air and light for any considerable time the delicate pink and cream color turns to a "murky brown," and the finish is in all respects most objectionable. The only natural and honest bleaching process known to the writer is that of "sumac baths." After the hemlock sides have been cleansed of all extraneous matter, as before described, by the most effective mechanical device known, it must then be hung in a vat of warm sumac liquor, and plunged frequently for one day (and even a few hours will sensibly affect the color). Usually one bag of Virginia sumac will suffice for a pack of one hundred sides. This process will cost about five dollars for a pack, or five cents per side, weighing fifteen to twenty pounds. The sumac liquor forms a vegetable acid, which acts most kindly on the grain of hemlock slaughter leather, not only removing (neutralizing) the color, but softening the grain, and contributes very much to the whiteness and clearness of the buff. Hemlock leather thus bleached will retain its improved color for a long time, and never go back to that muddy and objectionable color so common where other bleaching processes are employed. As it is the acid that effects the object sought, the sumac liquor should be retained long after its tannin has departed. As a mere tannin agent it is only valuable, as all goat and sheep skin tanners comprehend, while it is fresh, before the acid forms; but for the purpose of bleaching hemlock leather, it is questionable whether the old cast off sumac of the morocco dresser is not quite as valuable as new sumac. At all events, some experiments that have been tried go to this extent in their conclusions.

Slaughter hemlock leather, tanned with liquors of moderate strength, say 16 to 20 degrees, from the press leach, will come out with a color that is between the lemon and the orange; if to this we add the warm sumac process, we get a color so nearly a light lemon or a flesh color as to meet the requirements sought in the best oak leather. Indeed, for all fine work, whether men's or women's, the buff is superior to that of most pure oak tannages, for these have a "sickly white" which soils much more readily on the bottom than the flesh color of the hemlock or union tannages bleached as here indicated.

This bleaching process is particularly serviceable on calf, and all grain finished leathers, including harness and bridle. No purely hemlock tannage will "take the blacking" so well as leather which has undergone this treatment. With it, hemlock grain leather can be made to hold its color almost equal to that of pure oak tannage. Calfskins properly tanned in hemlock can, by this bath of warm sumac liquor, be made equal in color to the best French, German or Swiss; indeed, the resemblance is much greater than that of skins tanned by pure oak tannage, for the French color is controlled by the "larch" bark, the equivalent of our "spruce," which, as all know, is a modified hemlock in color. Above all these considerations actual experiments seem to indicate that this sumac process will add enough to the weight to pay for its cost. However this may be, it will add greatly to the intrinsic qualities of all upper stock and much to the beauty of the buff in sole leather.

As directly connected with this subject, it may be mentioned that for many years the union crop leather tanners used "sour milk" to wash the grain between the first and second rolling. This treatment not only "lightened up" the whole complexion, but removed clouded spots and even stains, and was withal a most harmless bleaching process, as the light and air did not affect unfavorably the buff, any more than in any of the vegetable processes. But mineral acids are objectionable and should be avoided.

CHAPTER XI.

THE CAUSES WHICH AFFECT COLOR AND ASSIST IN THE MAKING OF A VALUABLE EMBOSSING GRAIN.

WHY LEATHER SHOULD BE THOROUGHLY DRIED—STRUCTURE OF THE GRAIN—IMPORTANCE OF A PERFECT FINISH—CARE TO BE TAKEN TO AVOID STAINS AND DISCOLORATION—"CUIR" COLOR—THE NATURAL HEMLOCK COLOR—"RUSSIA LEATHER" COLOR—FRAUDS IN SELLING HEMLOCK FOR OAK LEATHER DURING THE WAR—COLORING TO BE DONE IN THE HANDLERS—EFFECT OF "STRIKING" THE GRAIN.

In the preceding chapter it has been insisted upon that leather should be thoroughly dried before being taken down to roll. The importance of great care in this matter may not be appreciated without some further attention to the peculiar nature of the "grain" and an examination of the causes which affect its structure, including its color and its embossing qualities. Some tanners, defending their practice, say, "Why occupy so much time in thoroughly drying the fiber of the leather, and then immediately thereafter dampening down again?" At the expense of seeming over nice, the writer will explain the reasons for taking this course.

The immediate outside grain of leather is a thin tissue, hardly thicker than thin paper. Next to this is an inner grain, several times thicker than the first, and very much more spongy in its nature. Together these two structures are usually called "the grain," and for the purposes of this discussion may be treated as one.

Much of the value of sole, harness, trunk and other leathers depends upon the appearance and condition of this grain. Upon this outer surface the harness maker and saddler stamp their forms and make their ornaments; the currier and finisher stamps or presses the grain with dies, making imitation hog, goat or seal skins out of ordinary neats leather. The boot and shoe manufacturer first buffs, then stamps and otherwise embellishes the bottom of his ladies' and gentlemen's fine work, on this grain, and the manufacturers of the finer leather fabrics emboss this surface with the most artistic forms. To enable these impressions to be made with proper effect and remain lasting—retaining permanently both form and color—the grain must be "clear," "bright" and "perfect" as to color, and "mellow," "elastic" and yet "firm" in structure.

This, then, is the nature of the surface or grain which is to be kept bright, and from which all coloring matter must have been washed and cleansed as far as possible by previous manipulation. If any device could be employed to pack the main fiber of the sole leather, and at the same time leave the grain perfectly smooth, it would be most desirable not to pack or compress this grain at all, but leave it soft and impressible for the artisan who comes after the tanner; but inasmuch as the inner fiber cannot be compressed by any known means without also compressing the grain, the next best thing to do is to so manage as to leave this grain as soft, elastic and compressible, and also as free from coloring matter as possible.

It is found by experience that the coloring matter of the bark, when in a soluble condition in the fiber, will spread, when compressed, almost as readily as ink on paper, and for this reason, after washing and cleansing the grain, it is found desirable to fasten the coloring matter in the fiber by evaporating all the water, which will leave the coloring matter in a dry, insoluble and fixed condition. This condition must not be disturbed—certainly not to the extent of making it soluble, by any after process of wetting; if that is done, the pressure of the roller will bring the coloring matter to the surface, and nullify all the advantages gained by the original washing and scrubbing, for the grain is almost as porous and as susceptible of receiving stain as blotting paper, and, when in a natural and proper condition, as impressible under heavy, as wax is under light pressure.

All that has or can be said of the proper treatment of leather to get rid of coloring matter, after it has been improperly placed in the fiber, should be subordinated to the better method of not allowing the coloring matter ever to go in the leather at all, or certainly not to the damaging extent which would render the extreme methods of mineral bleaching necessary. It is possible to extract the tannin from even our hemlock bark without so overcharging it with coloring matter as to damage the buffing qualities of the leather. It is not necessary that all our leather should be white, or cream color; any other color, if only natural and bright, is intrinsically as handsome and appreciable. It is only because the red color of the hemlock is thought to be extraneous that it is regarded as objectionable. It is because this color is thought to indicate our inferior quality or workmanship that causes leather having it to be condemned.

The writer saw, in one of the first boot and shoe manufacturers' show windows in London, the best English bend bottoms stained red, or in very close imitation of our hemlock. This was his trade mark, and had been for many years. There was no attempt at buffing, such as we appreciate in this country, but the bottoms were made perfectly smooth with the the "long stick." Now, this only proves that there is no standard color which, of itself, gives value. The bright hemlock color is now sought in pocket-books, satchels, ladies' belts, book-binders' leather—and, indeed, in all those leathers which go to make up the "art work," of which leather is the foundation. The French word "cuir," for leather, is to-day the name of the prevailing fashionable color—and this color is absolutely fabricated from hemlock bark; yet when it is found in sole leather it seems a badge of disfavor, to be got rid of by any means, however artificial and even damaging to the intrinsic quality of the structure of the fiber.

The fashionable "cuir" color can be most artistically made and preserved with pure hemlock tanning, if only too much heat is not used in extracting the strength of the bark. For fear that all may not understand what is the limit of heat permissible, it may be said that no tanner who desires to make the best color possible should use over 80 degrees of heat on his bark. Hemlock bark liquors, obtained with this limited heat, applied to pelts or hides properly prepared, entirely freed from all lime, will produce a color in almost exact imitation of the French "cuir" color, or, to put the expression into English, the "hemlock tan color."

The celebrated Russia leathers that enter so largely into the fine "Vienna leather goods" are originally of a light lemon color, produced by the willow bark with which they are tanned, and they are afterward changed by dyewoods and mordants into various colors; the most popular is that already indicated—indeed, it is sometimes called a "Russia leather color" by way of designation, and yet all Russia leather as seen in the arts is colored artificially.

A curious train of circumstances developed the impolicy of "hiding our true colors" during the recent rebellion. The army officers imbibed the prejudice which generally prevails that oak tanned leather was far superior to hemlock, and were influential in demanding this kind of tannage in all their army equipment work, such as shoes, belts, harness, etc. All of the earlier contracts specified that the leather should be made from "oak," and in some of the departments "white oak" tannage was called for. Soon the disparity between the price of hemlock and oak rough leather became so marked that the temptation to obtain in some way for hemlock leather the higher prices paid for oak became too great to resist, and curriers found a way to bleach the hemlock even to a lighter color than the natural oak, but by a process which greatly damaged the intrinsic quality of the leather. One of the first discoveries made of this fraud was the finding of a large lot of bayonet scabbard sheaths that had so rusted the steel which they were made to cover and hold as to make both leather and contents perfectly worthless. The writer was called on by the Government to determine the cause of this damage. The decision was that, by reason of the mineral salts used in bleaching, the dampness in the atmosphere collected an amount of moisture in the leather which caused the rust on the steel. Then how could this damage be prevented in the future? was the practical question to be determined. My suggestion was, to strike from the requisitions the demand for "oak tanned leather," and thus remove all temptation to make artificial oak; but this practical advice was not heeded, and, more or less, during the whole four years of our war, the Government paid for oak leather and got greatly damaged hemlock. This policy on the part of army officers cost the Government millions of dollars, and the regulations under which this mistake was committed are still, to a considerable extent, in force.

The *first* and *best* thing to do then is not to overcharge the tannin with the coloring matter by the use of heat, and second, if it is so overcharged, to get rid of it by washing and scrubbing and the use of vegetable acid, such as a warm sumac bath, for instance. A single liquor overcharged with coloring matter in the earlier stages of tanning will often leave its effect so permanently on the pack as to defy all after correction.

It is a mistake too often made by tanners to suppose that the coloring of the pack is effected by the last layers. Ordinarily the color given in the handlers is carried through to the end. Start the color right, create the proper mordant in the handlers, and then strong and even highly colored liquors may be used without causing serious damage. The writer once saw a tanner attempt to make union crop leather with the usual quantity of oak bark, by putting his oak in at the wrong time; he thought he should "finish off" with oak, whereas he should have used his oak as a mordant in his handlers and early layaways; the result was, as might have been expected—failure. There is good reason to think our English friends have made this mistake in using our hemlock extract. They have used the extract at too early a period, hoping to cover it up by their light coloring materials at the end—whereas, if they had colored their packs with terra and valonia, and put the hemlock extract in their last layers, they would have produced a different result.

The English custom of allowing their packs to "sweat" in piles before "striking out the grain" is founded on a sensible idea, and the practice even now of "striking" the grain rather than rolling the whole substance, is based on the practical wants of the trade. By this process the pelt is held in its natural state. The thickness is maintained, the edges (backs) are rubbed up and made to appear full, and thus a market value secured which would be sacrificed if the leather were dampened and rolled as is our custom.

Whether the manufacturers of Great Britain and Germany will ever become accustomed to our compressing process, and give us credit for plumpness which we seem not to have, is a question to be determined by future experience. If we change our rolling for a stamping process, and make our fiber hard, while at the same time we compress it, we certainly shall suit our German, Swiss and French customers, for this is the method followed by them in finishing sole leather.

CHAPTER XII.

CONSTRUCTION OF TANNERIES—THE TURRET DRYER.

HOW THE ADMISSION OF LIGHT AND AIR IS CONTROLLED IN THE TURRET DRYER—ITS CAPABILITIES FOR DRYING LEATHER IN QUICKER TIME, WITHOUT REGARD TO THE WEATHER—ITS CONSTRUCTION, AND HOW ITS CAPACITY SHOULD BE PROPORTIONED TO THAT OF THE YARD—HOW AND WHEN HEAT SHOULD BE USED—HOW TO PREVENT DISCOLORATION OF THE LEATHER—SAVING OF LABOR IN THE TURRET DRYER.

About the year 1864, at Sparrowbush, N. Y., the first turret dryer was erected. It was a six-story structure, with most of the improvements found in the present dryer. There is nothing new in drying leather in a tall building. Many of the old tanneries, built as long ago as 1830, had three lofts over the whole size of the tannery; the floors in the two upper lofts were latticed, and were therefore in this respect like the present turret form of dryer. But the principle claimed for this improved turret dryer is that, both as to air and heat, there is perfect control, whereas, with any drying loft heretofore in use, the damp air of the yard found its way up through the lofts, and in all cases there were opening windows from each loft, which were depended upon to admit the external air, so that substantially all the air obtained came in through these upper openings. The effect was that the leather hanging near the openings dried rapidly, while those sides hanging in the center of the loft remained unaffected, so that, before they could be dried, they were required to be transferred; often they were many times "shifted." This shifting process was particularly necessary in the winter time, when the large box stoves, burning wood, placed in different portions of the lower loft, or in the yard and beam house, caused the sides hanging near to dry rapidly and greatly to discolor, while at twenty feet distant the leather would be frozen solid.

The difficulties and great delays in drying sole leather by any of the old methods were among the most annoying incidents of the tanners' life. All these are entirely removed by the turret dryer. There is now an absolute certainty as to the time when the leather will be dry, and this does not in any necessary degree depend on the state of the weather.

Among the advantages of this new form of dryer are the following:

1. Drying in one-third the time—thus saving insurance and interest.

2. Drying without regard to the state of the weather—thus at all times keeping the rollers supplied.

3. Drying much more uniform in color.

4. Drying without shifting, or labor of any kind except to "hang up" and "take down."

These are some of the economies of the turret system of drying, but they might be enlarged upon. The principle claimed, as already stated, is the absolute control of heat, light and air; with these three elements under control, it will be admitted that all the advantages claimed must follow.

The building may be any number of stories high; some are but three or four, and others are seven and eight. Of course, to erect a building high enough to contain eight stories would require very heavy timbers, and from both observation and experience the writer would recommend but five stories, each about seven feet in the clear between beams —just high enough for a man to pass with his hat on. This structure need not be made of very heavy timber, and consequently would be inexpensive as compared with those buildings that run up so high. The question of convenience of elevating the leather, whether in the higher or lower turret, need not be taken into the account, as this is done by machinery. The building should be about two spans of timber wide—say forty feet—with two rows of posts, equidistant from the sides. Longitudinally with the two rows of posts should run a tight board partition, with intersections at every ten feet extending to the sides. This would cut the space up into two rows of rooms of about 10 by 12 feet each, with a center passage of about 13 feet. The roof is made in the usual latticed lantern form.

All the floors above the first or ground floor should be latticed, and the rooms would, of course, be immediately above each other, so that, if the building were five stories high, there would be five rooms 10 by 12 feet each, standing one above the other, and there should, of course, be just as many of these rooms, or series of rooms, as would be required to dry the stock of the yard, whatever its capacity might be.

Each one of these rooms will contain one pack of a hundred or one hundred and twenty sides, depending upon the weight of leather, and will dry the same in ten days. This would give the series of five rooms the capacity of drying fifty packs in ten days, or about fifty sides of heavy sole leather each day. Just so many times as fifty will go into the entire daily production of the yard will the tanner require duplicates of this series of rooms.

The rooms on the first floor should be supplied with steam

pipe, laid on the floor, or raised a few inches only by strips of hard wood, covered with hoop iron. The piping furnishing heat for each of these rooms should be under separate control, so that not only could the steam be turned on or off from each, but any *degree* of heat might be admitted.

All the packs taken out in one day, or in two or more days in succession, should go in one of these sections, so that the condition of the leather may be as nearly uniform as possible in each set of rooms. The leather is hung up on sticks in double rows, leaving a passage of nearly two feet between. For the first three or four days no steam should be allowed in the pipes or the section, nor should the trap doors which lead outwardly be opened but slightly if the weather is warm or the winds blow high, but in the fall, or when the weather is overcast, the lower trap doors may be safely left open. About the third or fourth day a very low degree of steam heat may be allowed in the pipes, and this may be gradually increased until the seventh to the tenth day, when it may certainly be premised that the leather will be fully dried; then all the sides in both tiers, and in all of the five rooms, will be dried about the same time, and may all be taken out and replaced by other sides. The writer has known turrets to be so actively worked as to turn out stock in seven days, but ten days' time is not too much, since it is very important not to hurry the drying the first few days.

It has not been stated that the center passageways should be lighted by cross sections leading to the windows, nor that each room should contain one small window close to the top, (but to admit light only), nor that there should be no openings except at the base of the lower room, with the air leading directly on the piping—all these are questions of detail that will be studied by any tanner who attempts to replace his old method by this new and better one—for often the general plan here outlined must be modified to meet the new conditions.

The "turret dryer" is beyond all question the most thorough and efficient method yet devised for drying leather, and in some of its modified forms should be adopted by all tanners. The difference in the temperature of the atmosphere at the ground and at an altitude of 40 or 50 feet would, of itself, create a draft, as is well illustrated by the erection of "stacks" or "chimneys" for the passing off of smoke or gas. But if to the natural action caused by the difference in the temperature we add a little steam heat, a steady, yet moderate circulation will be maintained from the bottom toward the top or openings of this structure, carrying upward and off the dampness of the leather, without creating such violent currents of air as to injure the color.

It is believed that the principle which underlies this method of drying sole leather could be applied with equal advantage to all other kinds of leather—calf, upper, harness, sheep, goat, and particularly such of these as are to be finished "fair," without injury to the color of the grain.

Returning to the construction of these turrets, it may be remarked that the center passageways will be found useful in affording room to run the trucks with green leather, or to store the dry leather awaiting the roller. Some tanners prefer to have shutes running through each loft, down which they slide the leather as soon as dry, to be piled on the lower floor. Others make their lattice floors of a temporary or movable structure, so that the whole contents of each room is dropped down into the lower room, including all the sticks on which the leather has been hanging. The writer does not think this a good plan, on the whole, since the extra labor of separating the leather from the mass of sticks, and carrying them back to their proper place, is about as much trouble as it would be to take the sides down in the rooms, and, with the aid of a suitable barrow or truck, run them to the slide which takes them to the dampening or storage room direct.

There is a most economical elevator or lift which should be known to all tanners who propose to adopt this turret dryer. It consists of an endless chain running from the extreme bottom to the top floor, ending under the roof. This chain should run in a wooden box, inclosed on three sides. The outward or open side will serve to attach the sides by means of hooks fixed to the links of the chain, say at distances of about four feet, depending on the rapidity with which the chain moves. The distances between the attaching hooks should be sufficient to enable the attendants to attach below and take off above the sides. By this economical method leather may be elevated to the highest loft with no more actual expense than if dried on the ground floor, and the sides, when once hung, remain until fully dried—thus saving all expense of "shifting," etc.

What has been said in another place upon the influence of light and heat upon color will measurably apply to this process. Leather dried in the open air will certainly dry dark, even if tanned with pure oak, and, if tanned with hemlock or a mixed bark, will darken to a damaging extent. If currents of air reach the leather while in a wet state, a like result is produced, with the addition of great harshness of grain. If a bright light, particularly if the sun's rays reach the grain or flesh, the leather turns brown, and is permanently discolored. The influence of the direct sun's rays, or even the strong light of the sun on vegetation, is a good illustration of such influences on the color of leather containing vegetable acid in solution. The ordinary table celery is covered with earth as fast as it comes to the surface, to keep the light from it, so that it may be *white* and *tender*.

Pie plant which grows under a barrel or in the shade will be white and not green. Grass that grows under cover, excluded from the light, is white, not green. This law of light applies to all vegetation. Availing ourselves of this principle, therefore, we say leather that is intended to be fair in color should be dried in the dark, and as free as possible from currents of air.

An illustration of the "turret" dryer, with further explanations of its construction and mode of operation, will be found in later pages.

CHAPTER XIII.

CONSTRUCTION OF TANNERIES — PLANS, FOUNDATIONS, ETC.

THOROUGH EXAMINATION OF PRESENT STRUCTURES AND APPLIANCES ADVISABLE BEFORE BUILDING—IMPORTANT CHANGES FROM THE PRESENT GENERAL USE OF STEAM INSTEAD OF WATER POWER—LOCATING ON "MANUFACTURING" AND "CULINARY" STREAMS—A LOAM, CLAY, OR SANDY FOUNDATION—FILLING IN BETWEEN VATS AND LEACHES WITH LOAM OR CLAY—PLACING THE VATS—THE "BUFFALO" VAT—"BOX" VATS—THE PROCESS OF "PUDDLING" IN SETTING THE VATS—UPPER CONDUCTORS—SIDE AND END WALLS.

As preliminary to all efforts to erect a tannery, drawings and working plans should be fully prepared. These plans should not be merely in the mind of the owner or builder, but they should be elaborately placed on paper, so that they can be well defined and susceptible of examination and discussion. The writer has seen so many expensive failures by reason of neglect in this respect, that, at the risk of being considered superserviceable, he would insist that, first of all, when the erection of a tannery is decided upon (and before the plans are drawn), extensive visitations should be made to the best constructed tanneries, where can be seen in practical use all the best-known improvements. The characteristic readiness, and even pleasure, with which American tanners show their works, and honestly discuss the merits of their methods, leaves no excuse for any man who proposes to erect a new yard not to avail himself of all the experience which has been obtained by others. On this trip of observation both the draftsman and the head mechanic should be of the company, and they should not stop short of visiting every tannery where a new idea can be obtained.

Leaving out of view at present the question of location for the economical supply of bark, hides and other material, we will treat at present only of the proper location of the buildings, irrespective of the markets for hides and leather, and the bark supply.

Until within the last ten or fifteen years, most of our tanneries were driven by water power, and this fact caused the buildings to be erected not only on the immediate bank, but usually the foundations were placed on the bed of a stream. This close proximity to the water was made necessary to enable advantage to be taken of the "head and fall," thus securing the greatest amount of power to drive the machinery. Such locations have always subjected the tanner to great danger from the overflowing stream, and as these risks were not insurable, many men have been entirely ruined by disastrous floods. But beyond these extraordinary risks, such locations were usually very expensive to prepare, since rocks and boulders had to be blasted and removed, and when this was done the foundations were often uneven and hard to adjust to the conductors and vats which rested upon them. Since the substitution of steam for water power, all these difficulties, and many more which might be enumerated, are avoided.

The tannery buildings should be located near a capacious and never-failing stream of water. The stream need not be large, for any direct purpose of the tannery, but a clear distinction should be drawn between "manufacturing" and "culinary" streams, as the health laws and public policy of all civilized nations make a wide difference between these classes of water courses. If a tanner locates himself on a "culinary" stream, he is always liable to the complaints of his neighbors below him—and even on a manufacturing stream he should, as far as possible, avoid throwing his waste in the water way, for, according to opinions of some of our courts, he is responsible for all *actual* damage done even on such streams. But if a tannery is properly located and constructed, it will be found profitable to retain and utilize all refuse animal and even vegetable matter, which so defile the stream when thrown in, and it really seems to the writer a providential and happy influence which intervenes to prevent the tanner from thus injuring himself.

The location of the tannery should, if practicable, be on a loamy, clay, or at worst a sandy foundation. If possible a loamy foundation should be secured. Whoever doubts that loam is equal to or even better, practically, than clay, should try some experiments similar to several which have come within my observation. Clay, if thoroughly worked, and "rammed" with great care, will, beyond all doubt, act as a good preservative of wood; but, by reason of the great amount of labor and care required in its manipulation, it often happens that the material is unequally worked, and spots of dry or unworked clay will be thrown in; this permits the access of air, causing defective parts in the wood, and soon the soundness of the whole structure is destroyed. Loam is much easier worked, and defects in its manipulation can hardly escape detection. The experiment which is commended to all doubters is as follows: Take a common pail and fill with loam, mixing and mingling water therewith, until the whole mass is of about the consistency of very thin mortar; then allow the contents to settle for a day or two, and the result will be that the water will stand on top and the earthy or loamy substance will fall to the bottom, in such order and compactness as absolutely to form a sandstone; indeed, it is just this process in nature that forms the sandstone which we everywhere see. When this loamy or earthy substance has fairly settled, holes may be bored in the bottom of the pail and the water will not percolate through, but will remain on top and finally evaporate.

The principle upon which this stone formation takes place must be observed in filling in between vats or leaches, otherwise there will be failure. When the whole mass is in a liquid or soluble condition, and is at rest, the settlement begins, according to the law of gravitation, the denser or heavier particles dropping first, and then the next heaviest, and so on until the whole body has settled just in the order of the specific gravity of its parts.

Now, it must be evident that, if the best result would be secured when puddling in between the vats or leaches, the whole mass should go in together and be plunged and mixed so that this order of settlement and adjustment may be the result. When loam is thus placed in between the vats it is almost impossible that there should be any leak. The writer has seen whole yards sunk without corking any of the joints, and yet all the vats remained tight.

When from any cause the foundation is defective—when gravel or projecting rocks are likely to allow water courses to be formed—the interstices must not only be filled, but made absolutely tight, and even new foundations must be artificially formed with loam, to the depth of one foot at least below the log conductors. If by any chance a water course should be formed under the yard, the entire profits of the tanner may run away. The tanner should remember that for a mistake in omitting to lay his foundations both deep and water tight, he is liable ever after to unconsciously waste his liquors. When it is considered that they are almost as valuable as malt liquors, he will comprehend the importance of the greatest possible care. No brewer would hazard the possibility of a leak in his underground tanks, however small, nor should a tanner.

When the foundation ground is thoroughly prepared — then, and not before, should the log conductors be placed. The vats may rest firmly on the conductors. The point of contact should of course be at the ends, where the tube connections are to be made. But in addition to the rest on this log conductor, timbers at least 6 by 8 inches should be thoroughly imbedded in the loam formation, not more than two feet apart, thus supporting the vats uniformly and equalizing the strain. Otherwise the vats will certainly leak after a short time.

When the conductors and supporting timbers are placed evenly over the whole yard surface, there should be a renewed attempt to puddle with a thin loam mixture—so thin that the smallest aperture or crevice underneath will be filled effectually.

At this stage of the construction there come in at least two different kinds of vat builders, with their plans of construction, both having merit. The first plan is known as the "Buffalo" method. This designation comes only from the fact that the Buffalo tanners first adopted it. It contemplates the foundations prepared as heretofore indicated; upon these foundations, plank, which with this plan should be at least three inches thick, are closely jointed and laid over the whole surface, and then spiked to the timbers. Sometimes these plank are tongued and grooved, sometimes only jointed up close, and when the plank are half seasoned the joints may be trusted to close by the action of the dampness swelling them tight; but a corking joint, well filled with oakum, is, in my judgment, much the safest reliance. After the flooring plank have thus all been evenly and permanently laid and corked, the grooving plane cuts the grooves, into which the planks of the vats are inserted endwise, so that, when the vat is formed by these prepared plank standing endwise, resting in these close-fitting grooves, each side and end of the vat is keyed up by a "wedge plank."

The only advantage that this form of vat has over what is termed the "box" form is that there is no space wasted between the vats, and, of course, if leaks occur, they can only be from one vat to another. In locations like our large towns and cities, where land is very valuable, this method of constructing vats has generally prevailed, as it saves at least four inches space between them. The writer cannot, however, but think that, both on the ground of economy of construction as well as safety, the box form is to be preferred.

The ordinary "battened" and "box" vat is, in my judgment, the best, because the safer form, where lumber is cheap and where space is of no value, as is generally the case in new territory, where our tanneries are usually constructed. This form of making the vats may be thus described. After the preparation of the foundations, as heretofore indicated, boxes made of plank, either two or three inches thick, battened together, are placed side by side and end to end over the whole surface. The battens are "dropped in" and "pass each other;" or, to use the mechanical term, "joints are broken," so that the battens form supports to both sides and ends of the vats that come into contact and adjoin each other. When thus placed, connections are made, by tubes through the bottoms, with the log conductors beneath, and the seams are thoroughly corked with "spun" oakum. The best form of making these seams, and the most reliable and economical method of making tube connections, it is impossible in words to communicate, but it may be assumed that whoever really intends to avail himself of these suggestions will think it worth while to investigate more fully before he attempts to put in practice ideas which, at best, are very imperfectly described in these chapters. There are economic methods of preparing the plank for these vats, which have been employed within a few years, which put to shame the older hand methods. Whether the plank for vats should be made of pine or hemlock timber, and whether from plank two or more inches in thickness, are matters of detail, which it would be foreign to the object of this chapter to consider.

After these box vats have been placed, corked, and the proper tube connections made, then comes the important work of puddling. Many yards have been ruined for want of care in this particular. The vats must be filled with water just as fast as, and no faster than, they are "puddled in" from the sides and ends. Some tanners think it sufficient to hold the vats in place by putting weights on the top, or by studding from the floor upward, but neither of these forms should be relied upon, since the pressure of the concrete, or puddling, from below, is very great, and it should be met by an equal, uniform, downward pressure—such as only the weight of the water filling can give. The loam which is used to fill in between the vats should be placed convenient to the yard, should be prepared in tight mortar boxes before it is run in, and so thoroughly mixed that it will run in box shutes to any desired spot. Carrying in pails, or otherwise handling, makes slow work, and too often induces the throwing in of the dry loam and the attempt to mix by plunging. This form leads to air holes and imperfect puddling. The whole process should be carried through on a uniform system, and no slighting or imperfect work should be permitted.

Mr. James Clewer, who was the author of this system,

always claimed that it was possible to so perfectly puddle vats with loam as to render corking unnecessary. Indeed, he did establish two or three yards on this plan without any corking whatever, and to this day (now more than thirty years afterward) the yards do not leak outwardly, though some of the seams leak inwardly—that is, the drip and draining of the loam finds its way through the seams, and discolors slightly, but only occasionally, the sides that come in contact with it.

By omitting the precaution of filling the vats with water before puddling, the result will be not only that the vats will be raised from their foundations, thereby disturbing the tube connections, but the sides and ends will be forced inwardly, so that ever after the shape of the vats will be distorted and their capacity lessened.

The "caps" or alley flooring resting on the tops of the vats should be often raised to see that the filling has not given way. If, as is quite likely for the first few months, such giving way or sinking is noticed, great care should be observed in refilling. When once the loam has thoroughly settled and become fixed in place, it may be considered certain that a perfectly tight yard is guaranteed for all time.

Some tanners have an upper system of conductors through which they supply their vats with new and strong liquors. These, when new and perfect, are very convenient, but such conductors are not to be trusted. After a few years they become decayed and leak, to the great damage and waste of the liquor. No care in the placing seems adequate to keep the air from these wooden structures when made so near the surface, and when the air does reach the wood it is certain to rot it in four or five years. It is for this reason the use of these upper conductors between the vats cannot be recommended, even where they are partially covered with earth.

It is far better to have the upper conductors wholly above the top of the yard, either in open shutes or in tight log conductors, where any leak can be detected.

When a tannery is located on an earth foundation, the side and end walls need not be extended to the lower foundations. Excavations can be made, the yard placed and filled in, and side and end walls may be commenced on timbers laid within two or three feet of the surface. If the timber is laid below the frost, and below the point reached by the air, it will last forever, and it will be safe to lay a brick, stone or concrete wall from this timber foundation upward, say two or three feet above the surface of the ground, on which the sills of the building may finally rest, beyond the reach of the damp earth beneath. If the filling in on the outside of the vats has been thoroughly done, this filling gives a more secure foundation than the natural earth, and will save much expense in the foundations.

If it is thought desirable, as it is in the opinion of most tanners, to erect the frame before setting the vats, the frame can be supported by temporary posts, running down to the foundation, and the making of a more permanent foundation may safely be left till the filling in is done, as here indicated. It will be useful for all tanners to remember that hemlock is just as lasting *under ground* as pine or other wood.

CHAPTER XIV.

CONSTRUCTION OF TANNERIES—LEACHES.

ROUND OR SQUARE LEACHES—THE DURATION OF LEACHES ABOVE AND SUNK IN THE GROUND—HOW TO BUILD A ROUND LEACH—HOW TO MAKE AND SET LEACHES IN THE GROUND—THE CAPACITY OF THE SETS OF LEACHES TO BE PROPORTIONED TO THE SIZE OF THE TANNERY.

The form and construction of leaches is a problem of great importance to tanners. Shall they be round or square? Shall they be above or underground? Shall they be constructed of wood, stone, or brick, and, if the former, then shall oak, hemlock or pine be used? How can they be most economically constructed? How long will they last made of either of the materials or forms named? All these questions, and others besides, will occur to the practical tanner who contemplates building a tannery; they are included in the general question, What kind of leaches are the most economical for tanners' use?

Since the introduction of the "sprinkler leach" most tanners have made their leaches round. It is quite a mistake to suppose that the patented improvement covers the use of any kind of a round leach. It is true, probably, that a round leach is better adapted to the "sprinkler" than any other form, and for this reason the patentees have adopted it. There certainly is nothing novel in this form since, on the

Continent of Europe, all tanners use round vats, and leaches, too, whenever they use leaches at all. If, therefore, tanners conclude that round leaches are preferable to square ones, they are quite at liberty to use them without any patent claim.

It is safe to assume that no leach made of wood, however constructed, will, if placed above ground, where air has access to it, last more than four or five years; by that time leaks will become so plenty as to necessitate a renewal. It is said of a chain that the whole is no stronger than its weakest link; so it may be said of a wooden tank or leach—the whole is no more lasting than its most imperfect stave or joint. Practically, then, whenever a single defect occurs the whole leach must be abandoned. It would never be wise to put new staves in a defective leach, any more than the housewife would think of mending an old water-pail by replacing a defective chime.

In considering, then, the economy of round leaches placed above ground, we are to estimate the duration of one of these structures by its single parts, and not by its whole structure. The experience of the most successful tanner using these round leaches above ground will show that four, or at most five, years is the duration or life of one of these structures. Reference is here made to the *tight* round leach—a leach that can safely stand full of strong liquor without "receiving leaches" or "drips" standing underneath, for in this manner these leaches have always been constructed and worked. But if a tight plank flooring should be constructed, and the leaches placed above it, the writer never could understand why attention should be paid to a small leak; indeed, if the leach was never filled with liquor, why should there be any pressure on the joints? Why should there be any leak, or if there was, why should any loss or injury result? This assumes that the bark is percolated by the liquor as fast as the latter passes the sprinkler; that the liquor never floods the bark—and, so far as the principle of the patented improvement is concerned, this form of operation would more fully answer the purposes of the patent than if a portion of the bark was flooded—that is, had standing liquor in the leach. Indeed, just to the extent that liquor stands in the leach the patented idea is lost sight of. It then becomes a press, rather than a percolating or sprinkler leach. The form of running these leaches should properly influence their construction, for there is no reason why a sprinkler leach, run upon the true principles of the patent, should ever rot or wear out, since, so long as the staves stand up or the bottom remains in, it will inclose and hold bark; and, so long as the staves would hold together to guide and confine the liquor in its downward course, a sprinkler leach would still remain to perform its office. Of course this defective structure would necessitate a tight plank flooring beneath; but, so far as the writer is informed, the patentee has never recommended this form of handling these leaches, and until he does we must treat them as tight leaches, and subject to be renewed once in four or five years, under the most favorable circumstances, when they stand above ground.

In deciding upon the material to be used in building leaches, it should be understood that young, sound hemlock is just as likely to last the allotted time as the best pine; but care should be taken to have the staves made from uniformly young trees, perfectly sound and fresh—that is, just peeled and sawed, and not from either old timber or that which has been lying two or three years before being sawed. Young hemlock trees, say not more than one foot in diameter, sawed into staves six inches wide and two inches thick, can be manufactured into tank or leach staves almost as fast as they can be picked up and passed twice through the hands of the employe.

The writer gives a little detailed instruction upon this point, since, through unreflecting employes, he was once made to suffer many thousands of dollars' loss for the want of a proper knowledge of the economies of this subject. After the plank havè been sawed of the requisite length and width—usually seven feet long and six inches wide—the staves in this rough form are beveled by a circular saw on both edges. This bevel, of course, is made to adjust to a circle of 8, 12 or 16 feet, depending upon the diameter of the leach. When the staves are thus uniformly beveled, they are laid in a circular form, ten or fifteen at a time, and are chimed out by a "chiming" or "grooving plane." After being thus formed they can be set up around the round bottom into which they are driven, and where they are held together by iron hoops, without calking; such a leach will certainly be tight.

The saw that makes the bevels should be about eight inches in diameter, and be very straight, fine and even in its set, so that a perfectly uniform surface may be secured. It is said "uniform," rather than "smooth," because it has been often demonstrated that such a surface, when pressed together will form a tighter joint than can be secured by the hand plane, however much pains may be taken to secure smoothness. Very great attention should be paid to the character of the material. Hemlock is so cheap and abundant that none need be taken that is not perfect, both as to soundness and uniformity of thickness.

In making the bottom of the leach, hemlock plank, plump two inches thick, should be used if the leach is to be eight or twelve feet in diameter, and if it is to be sixteen feet in diameter then the plank should be three inches in thickness;

9

in both cases the edges should be chamfered down to a uniform thickness of two inches "scant." On the supposition that the leach is to be made twelve feet in diameter, the plank may be sawed sixteen feet long, when they will cut into bottom plank to good advantage. But it is seldom worth while to attempt to economize so closely as to save the whole of hemlock timber, and therefore, at the expense of a little possible waste, only perfect plank should be used in the bottom. These, after being square-edged with a small fine-set circular saw, should be laid down on a platform bench, and firmly griped together, secured by a temporary batten; when thus placed a circle should be inscribed the size of the proposed leach. With a small whip saw this line should be very carefully followed, taking great care to make the cut square, and at a right angle with the surface of the plank. If this is done with care and with a suitably straight saw, there need be but little after-work with the plane to make the circle smooth and otherwise perfect. If care has also been taken to have the plank originally sawed of uniform thickness, a very little work with the plane will bring the edges to a uniform thickness, so that when put together the outer rim of the bottom will exactly fill and fit the chime groove in the stave, making a reasonably good, tight joint.

We are now ready to set up our leach, and we want round hoops of at least five-eighths or three-quarters iron. The lengths should, as far as possible, be purchased to suit the size of the leach. Usually this round iron comes twelve feet long, but it may be obtained of any length desired. There should be but one joint for each hoop—a sufficient length of iron being welded together to go completely round the leach—and the ends should come together in a cotterel, which may be of cast iron or hard wood. If one of the ends of the iron should have a screw cut on it for six inches, while the other is firmly fastened in the cotterel, a nut and wrench will draw the hoop as tight as it could be driven, if the form of the leach was tapering in shape, as on a pail or ordinary wash tub; but as it is desirable to have these leaches of uniform diameter, this form of drawing together and holding the stave must be resorted to.

There is nothing difficult or expensive in constructing this form of leach, if the proper skill and machinery are at hand, but without this machinery and this knowledge they are very expensive, and often fail of being tight. With the small saw and mandrel, which cost about thirty dollars, and a chiming plane, which will cost about eight dollars, these round leaches may be put together with ordinary carpenters' tools at a very inconsiderable cost. The cost of the labor need not be over ten dollars for an ordinary leach, twelve feet in diameter, and seven feet deep in the clear. The expense may be increased by a greater regard for permanency in its construction, as the material may be clear pine or oak, the hoops may be of heavier iron, and instead of four hoops (the usual number) there may be six. But in this, as in all tannery construction, as much simplicity and cheapness should be observed as is compatible with the service to be performed. The iron hoops used will last during the lifetime of several wooden structures. On the whole there can be no doubt that round leaches, constructed as here suggested, should always be employed when the leaches are to be used above ground; but for underground leaches, or leaches filled in with earth, quite a different construction is suggested.

Until the introduction of the Allen & Warren sprinkler leach, square leaches set in the ground, in a packing of loam or clay, were in general use, and to-day are not abandoned even by our best tanners. A set of these leaches, properly placed and filled in, will last for twenty or thirty years; indeed, with slight repairs to the top planks, they will last during the lifetime of a tannery. The writer knows of several sets that are now in good order that have, with slight repairs, been in active use for more than a quarter of a century.

The preparation of the ground on which these leaches stand should be made with even more thoroughness, if possible, than that of the yard itself, since there is more weight resting on it, and more disturbance from the flow of currents of water. If the leaches are constructed with proper openings in the bottom, through which the spent tan is washed after the leaching process is over, then more than ordinary care should be observed in placing the under conductors, otherwise air will reach the plank flooring and cause it to decay. In all locations where a flow of water can be secured on the top of the leaches, provision should be made for this to wash out such portion of the spent tan as may not be wanted for the furnaces. Sometimes, however, the nature of the stream is such that tanners are not permitted to throw their refuse tan in the water way, and in such cases this economical arrangement cannot be availed of.

Ordinarily these square sunken leaches are of 10 by 12, 12 by 14, or, in exceptional cases, 16 by 20 feet, surface measurement, and usually about seven feet deep. If the leaches are of the smaller sizes, then planks two inches thick are quite sufficient; but if the larger size, then planks at least three inches thick should be used; not more, however, for the economical reason than from the fact that the press can be better controlled in the smaller sizes, does the writer commend this form of leach. Indeed, the number and size of leaches should correspond to the size of the vats in the yard. The covering of one leach of bark should make one vat of liquor. There should certainly be no fractions. It should either be one, two, three or four vats of liquor. Attention to this seeming detail will much simplify the manipulations afterward.

The preparation of the plank for these leaches can be made by the same saw and mandrel which has been commended for the round leaches; even the calking seam can be formed with this saw. No joint can be made with a hand plane that will hold oakum so firmly as with this saw joint, and the oakum joint is made much truer by the saw than it can be made with hand labor.

When the ends, sides and bottoms have been battened properly, they are put together as in the case of the vats described in a previous article. Indeed, these leaches are but so many vats enlarged in size.

In Great Britain the tanners make use very largely of brick and cement to form their vats and leaches. No doubt these materials make a very substantial structure, and as erected there, in cities and towns, and intended to last for all time, this construction may be the best. But some attempts made in this country to use this material have resulted in staining the leather. Whether this difficulty could not be overcome with us, as the English tanners claim it is with them, it seems hardly worth while to consider, since timber is so much cheaper here than brick, particularly in the country, where all our tanneries are built. Wood plank will last fifty years, sunk in our usual way, and that is quite as long as our civilization will tolerate the existence of a tannery in one location.

CHAPTER XV.

CONSTRUCTION OF TANNERIES—FRAME WORK AND LOCATION OF BUILDINGS.

WHY THEY SHOULD BE ONLY ONE-STORY HIGH FOR THE YARD AND BEAM HOUSE—SAVING IN INSURANCE BY SEPARATING THE BUILDINGS—CONVEYING LEATHER TO THE "TURRET" DRYER—TRANSMITTING POWER TO DISTANT BUILDINGS—PROPER SPEED FOR BARK MILLS AND ELEVATORS—SIMPLE PROVISIONS AGAINST FIRE AND BREAKAGE, AND TO PREVENT DUST.

A modern tannery is quite a different affair from one of the early Greene County (N. Y.) structures. The various improved methods of heating and obtaining power have rendered it quite unnecessary to crowd into one building, as formerly, bark mills, rollers, hide mills, drying lofts, yard, beam house, sweat pit, etc. Heat and power can now be used without limit, and wherever their use can be made to economize labor or cheapen insurance they should be employed.

That great advantages are secured by a one-story structure over the yard and beam house will be conceded when the following points are considered:

1. The timber need only be heavy enough to carry and uphold the roof, not forgetting, of course, its probable load of snow in winter.

2. No apprehension need be felt about the falling in of the structure for a long time after the usual decay of the timbers

commences, since they can be replaced without inconvenience, there being no heavy superstructure to sustain, whereas a building with lofts above, often filled with wet leather, is always an object of solicitude to the tanner; he is constantly studding up and "supporting" the building to make it safe for the workmen. Besides, these high structures are subject to the action of the wind, which much weakens their joints and fastenings.

3. The insurance is only one half the price of what was paid on the old and high buildings. Although these one-story structures have now been in use for ten years or more, several hundred of them being in existence, not a single one has yet been destroyed by fire, and if this record shall not be changed by further experience, we may reasonably expect a still further reduction in the rate of insurance. In fact, with an abundant water supply and efficient service, a one-story tannery cannot be wholly destroyed, for nothing but the roof could burn, and as this can be reached from both above and below with ordinary water buckets, the progress of the flames can be stayed in almost any case of fire likely to happen. Besides, the condensation of water from the steam of the yard always keeps the roof boards and covering water soaked, so that fire would not spread rapidly, if at all.

4. A building of one story for the yard gives opportunity for a high ceiling and good ventilation. It has, however, this single disadvantage—that in winter it is much more exposed to the frost, although, practically, ice seldom forms in a yard of this construction. Indeed, experience has shown that with a few coils of steam pipe running around above the sills there need be no cold fingers of the workmen. This one-story structure should have a flat roof (one foot pitch in ten) covered first with boards and then with asphalt paper and gravel, by a process universally in use.

So far, then, as the yard and beam house are concerned, a one-story building is the most suitable structure. But how shall the tanned packs be got to the drying loft? Heretofore trap doors have been placed in each bent of the building, so that little more was needed than to open the trap and hook the sides from the yard below, and thus pass them up by hand from loft to loft. This seeming convenience prevented any change from the old method for many years. Tanners not unnaturally reasoned in this way: "We have a yard and beam house, and they must be covered with a roof; the same roof can equally cover our drying lofts, and as our drying lofts must have a capacity as large as our whole beam house and yard, in order to dry our stock, why should we not make our structure strong enough to carry all this under one roof?" Besides, they have reasoned: "Our insurance covers our stock in both yard and loft, and at the same price; why should we go to the expense of erecting a separate structure, where our insurance would be divided, without any compensating advantages in the way of reduction of rates?" Thus reasoning, our tanners went on from year to year erecting new yards under former plans, until within the past ten years, during which many have been induced to break away from this old method, and are now building turret dryers in connection with the rolling or finishing lofts, connecting these with the yard by tramways, hung from above in some instances, and in others with bottom rails of wood, these tramways running cars through the center of the yard, extending out to the drying lofts. The form of erecting these tramways varies with each location. Where a tannery is situated on level ground, accessible on all sides by a horse and truck, probably there is not a more economical way of transferring the wet stock from the yard than by this means. In this case the tanner can have openings from his yard at frequent intervals, and the nearness and saving of transportation of the wet leather in the yard will compensate for the cost of the service of the horse and cart, over and above a tramway through the center of the yard, to which all tanned packs must be brought.

The latest and most approved method of locating the buildings so entirely separates the yard from the drying lofts that the fire risks are greatly reduced, for the yard and beam house are one risk, the leaches and bark mill another, and the furnaces and boilers still another. The building containing the latter should be placed at least 100 feet from all other structures, and in itself made perfectly fireproof—that is, there should be no wood anywhere near the structure containing the furnaces and boilers. So arranged, it is found practicable to run steam power off in any direction and almost to any distance. The most economical way of running off this power is by means of steam pipe, thoroughly protected with ashes, loam or clay, well packed in a box surrounding and inclosing the steam pipe. The engines that drive the machinery may be in distant buildings. Some tanners drive their machinery at a distance of from 300 to 1,000 feet from the boilers, and the condensation does not seriously affect the power, a portion of which is necessarily lost, but when wet spent tan is burned this loss has no commercial value.

It has been demonstrated that steam power can be conveyed in pipes more economically than by running shafting; that is, wherever power is required a steam engine should be placed and steam conveyed to it, rather than to depend on one central engine, and either "belt" or "shaft" off. Engines are now constructed so strong and cheap, and to run with so little attention, that one engineer can take care of two or three with the same facility as one. The steam valves are opened in the morning, and are not shut or otherwise disturbed until noon, and then again are opened at 1 P. M., and not closed until sundown. So different is this practice from the old method, and the care made necessary by the defective character of steam engines, as formerly constructed, that engines may now be multiplied to any required extent without making it necessary to employ an assistant engineer.

It may be safely estimated that a one-story yard and beam house, isolated at least 100 feet, can be insured for 1½ per cent., while 3 and even 4 per cent. is now charged on ordinary tanneries constructed after the old methods; but no insurance is needed on the leather in the vats, for neither the leather nor the vats themselves will burn, even if the framework above is destroyed.

The turret drying loft (of which mention has been sufficiently made in a previous chapter) should be near the yard, and should be classed in the same risk; but the leach house and bark mills should be in an opposite direction, and as far as possible from the central or boiler house and furnaces. Where the ground will permit, the boilers and furnaces should be in the center, the leach house and bark mills fully one hundred feet to one side, and the yard and turret drying lofts as great a distance in the opposite direction. This will bring the important structures two hundred feet apart, and if the boiler and furnace building is built as it should be, of stone or brick, and is not over fifteen or twenty feet high, there is really an unobstructed space of two hundred feet between the main fire risks. There is really no manufacturing structure in the country so free from accidents by fire, when thus constructed and situated, as one of these one-story yards and beam houses; and the drying lofts, always filled in part with wet or only partially dry leather, have a moist or damp atmosphere which, with steam pipe only used as heaters, renders anything like accident from fire almost impossible. The natural, and, indeed, the only fire risk about a tannery, comes from the bark mills and bark elevators, and in a vast majority of cases the fires which consume so many tanneries proceed from this source. How can this risk be lessened or altogether avoided? The bark mills and elevators should have a slow motion. No mill should run above sixty to eighty revolutions per minute, and the elevator belt should run slow enough to enable the eye to take in and count the buckets or boxes as they pass a given point. The shovers and screen should move slowly, for besides the friction which causes fires there is much more wear and tear to a quick than to a slow motion, and power is also lost.

An improvement has of late been suggested to suppress the dust arising from the fine ground bark, as follows: A steam pipe, half an inch in diameter, connected with the nearest direct pipe from the boiler, is brought to the underside of the mill; a very small jet of steam is allowed to escape, to dampen by condensation the bark as it comes from the lower throat of the mill, and before it drops into the conveyers. This steam pipe could be so placed that if fire should occur in the elevators, from friction or otherwise, a full head could be turned on by the attendant, and steam forced upward through the entire length of the elevator box. This, therefore, may be regarded as an inexpensive contrivance to prevent dust, and also as a great safeguard against fire; and now that steam is so cheap and abundant there seems no reason why it should not be adopted by all tanners, especially as it is regarded by insurance companies with great favor. Bark elevators are liable to get choked and stop; when this occurs the bark accumulates and gets packed in the lower part of the mill; great friction and danger from fire arises from this source, and to avoid the danger an "overflow" or space under the mill should be provided. But a "tell tale" should be put on the elevators, plainly in the sight of the bark grinder, so that he can see when there is a stoppage. Some tanners think it a sufficient precaution to have an opening in the elevator for this observation; but a "tell tale" should also be placed on the elevators, and this small and inexpensive lever should, in its rising and falling motion, indicate its action by a slight noise, so that if the eye is otherwise directed the ear of the attendant will notify him of the danger, which is from breakage as well as fire.

CHAPTER XVI.

THE ROSSING OF BARK.

THEORIES OF THOSE WHO ADVOCATE ROSSING—ITS COST—DIFFICULTY OF ROSSING WITHOUT TOO GREAT LOSS OF TANNIN—STRENGTH OF LIQUORS WHICH MAY BE OBTAINED FROM ROSSED AND UNROSSED BARK—POSSIBLE ADVANTAGE IN ROSSING BARK FOR EXPORT IN THE "LEAF."

To what extent, and under what circumstances, should bark be rossed before being ground for tanners' use? If credit be given to the statements of parties interested in the manufacture and sale of rossing machines, tanners will be led to the conclusion that rossing is an absolute economic necessity. As the writer does not agree with this sweeping conclusion, and yet believes that there are circumstances under which it is profitable to ross bark, the limitations which should govern in this matter will be briefly considered.

The theory on which the rossing of bark is advocated may be thus stated:

1. The outer ross, or dead bark of the tree, contains no tannin. By the exposure of its weather-beaten surface to the rains and winds, all but the woody, fibrous structure has been destroyed. To place this porous, spongy, woody fiber in contact, as when ground, together with the extractive and tannin matter of the live portions of the bark, which are charged with tannin, is to absorb and dissipate this valuable

product to no purpose. In fewer words, this woody fiber will thus be *tanned*, and this it is claimed is wasteful.

2. The space which this ross occupies in the leach displaces so much good bark, and to this extent retards the leaching by lessening the capacity of the leaches.

3. This dry ross is a valuable fuel, and can be used to some advantage in the generating of steam or for other purposes.

Against these advantages should be placed the cost of rossing, which is quite inconsiderable, if the advocates of rossing are to be credited. Conceding all that is claimed for this practice, it is impossible by any machine yet invented, or even by hand labor, to separate the worthless ross from the live bark so exactly as to make the process profitable for ordinary tanners, who buy and use their bark at home in the interior locations. Let any tanner who doubts this statement examine carefully the ross coming from any rossing machine. Let him place a limited quantity under the influence of hot water, and the tan liquor present will be a convincing proof of the inexpediency of the practice. Or let him take a piece of bark no more than one foot square, weigh it carefully, and then, by hand process, with the aid of a sharp knife, let him attempt to remove just the colorless dead ross, and no more, taking great care not to touch the live bark. After this is done contrast the percentage of rejected ross with the percentage obtained by means of the rossing machines, and the result will show just how much live bark is taken for ross, to burn up, by the use of the machine. This latter experiment, conducted never so carefully, will also prove how impossible it is to divide exactly and positively the dead and worthless from the live and valuable bark. If it cannot be done by hand, even in a small experimental way, how much more impossible is it to do with a machine which cuts the thin and thick bark alike—notwithstanding all attempts to adjust the knives and rollers to suit the substance of the bark.

The testimony of tanners differs as to the percentage of ross removed; most of them, however, agree that even one-fifth, or 20 per cent., is about the average loss. Some chemical tests of the comparative strength of rossed and unrossed bark have been made, and, naturally enough, the percentage of tannin in the rossed bark has been found in excess of that in the unrossed. But a recent and a more satisfactory experiment gave the following results: A piece of average hemlock bark was weighed—we will suppose the weight to have been sixteen pounds. This piece was first cut evenly in two parts, and both parts were afterward made to weigh exactly alike. The ross was carefully removed from the one piece by a hand knife; (only about one-tenth of the weight was removed, showing the care with which it was done.) Both pieces were then separately ground fine, and leached, great care being observed to continue the equal condition all through the experiment. The result was, by the barkometer test, the extract obtained from the unrossed portion stood higher, that is, indicated a greater degree of tannin, than that portion which had been rossed. This is an experiment which any tanner can try in a few hours, and at small cost of time and labor. If the experiment is tried accurately, it will be found that, without considering the cost of rossing, there is a positive economic loss in the process. In no instance, however, has the strength differed equal to the loss of the ross, which we assume to be 20 to 25 per cent. The experiments show that hemlock bark that is rossed will give 8.66 per cent., while the same bark not rossed will give 7.13 per cent. The experiment was made by a celebrated Boston chemist in the interest of the patentee of a bark rosser, and may at least be taken as the most favorable result which can be produced.

If this is the result of the comparative strength of bark, rossed and unrossed, we are left to consider on the one hand the cost of rossing, and on the other the value of the dry material. The cost of rossing, independent of grinding, may fairly be placed at fifty cents per cord, and the dry ross obtained equal to 400 pounds. Is 400 pounds of ross worth fifty cents? Of course this would depend upon the value of fuel in the locality. To the ordinary tanner the dry ross is worthless, and would hardly pay for its removal, since the wet spent tan furnishes far more steam power than suffices for driving the whole machinery of the tannery, and dry ross is a poor substitute for wood anywhere else than under a furnace.

But there are circumstances which would justify rossing as, for instance, if bark had to be transported long distances, where the cost of transportation was a considerable item of the value; or, if the bark had to be "baled," and the freight was based both upon bulk and weight, then would it be especially necessary to ross. It has often occured to the writer that both the English and German tanners would find it profitable to buy our oak and hemlock bark and closely ross and bale for shipment to their market. Rossed bark in the "leaf" will occupy one-quarter less space than when "in the rough," and with care may be so compactly placed in a bale as to make solid stowage—so solid, indeed, that no ordinary leverage can compress it. When thus baled, and bound with wire, there can be no doubt that bark could be sent to Europe from many of our Southern ports to advantage. The cost would be as follows:

Original cost of oak bark, per ton......	$10 00
Cost of freight to seaboard............	4 00
Cost of baling and rossing.............	2 00
Freight to Liverpool..................	5 00
Total..........................	$21 00

Southern oak bark, thus rossed and baled, would give European tanners better and cheaper tanning material than they at present employ from any barks in the "leaf" or "chip" which they now use. Whether "extract" from this bark could not more economically be sent abroad, is quite a different question, which the writer will not here enter upon.

CHAPTER XVII.

UTILIZATION OF TANNERY REFUSE.

BURNING THE WET TAN—GLUE STOCK—IMPORTANCE OF KEEPING THE PIECES PURE AND SWEET—PRESERVING, CLEANSING AND DRYING THEM—USES FOR CATTLE HAIR—THAT WHICH COMES FROM SWEAT OR LIMED STOCK—WASHING, DRYING AND PACKING—FERTILIZING LIQUIDS FROM THE LIMES AND SOAKS.

Previous to the year 1852 it was customary in all the tanneries of New England to dry, or partially dry, in the open air, all the spent tan from the yard and leaches, and store during the summer months for winter use, not only for heating the liquors, but, in exceptional cases, to generate steam to furnish the power for the ordinary work of the tannery. Even up to this day the practice is regarded as most economical by the upper leather tanners, and such others as do much finishing. None of the patented, or unpatented, methods of burning wet tan can secure as good a result with wet as they can with dry tan. Experience has shown that from one-third to one-half the effective heat-producing power is neutralized in driving off the water. In other words, there is no method of burning wet tan that does not first evaporate the water. Therefore, when there is an inadequate supply of tan, and it is convenient to dry it, it is always better to do so. But in the case of our sole leather tanners, they have a surplus of wet tan, and to waste it by burning a profuse quantity is to save the expense of otherwise disposing of it.

In the year 1852 Mr. Joseph B. Hoyt conceived the idea of burning wet spent tan in a detached brick furnace, and did for the first time in America succeed in obtaining the power to drive his machinery by this agency, unaided and alone. This was done at his tannery at Woodstock, Ulster County, N. Y., and the event has been made memorable by the great success which has attended the practice ever since. This improvement revolutionized the construction of our tanneries. It rendered water power of little or no value. It caused tanneries to be located upon open flats, where access could be had from all parts to the buildings, and took them from the banks of the streams, where floods and overflows did yearly great damage; it rendered unnecessary the use of wood or coal for the heating of liquors, or the warming of the lofts, and, by the unlimited and inexpensive use of this newly-acquired agent, labor saving machinery was introduced everywhere.

The controversy about the priority of the invention or use of wet spent tan, commenced in 1869, occupied the attention of one of our courts for six years or more, and the decision finally reached was that Mr. Hoyt and his agent, Mr. Crockett, did introduce these wet tan burning ovens as early as 1852, and by their use a result was obtained equal to any which has been secured by any improvement since made.

Notwithstanding the publicity which has been given to the construction of furnaces for the utilization of this refuse, there has been no one specific plan that has received general commendation. So many of these furnaces have been adapted to the condition of the structures already in existence, that, with the exception of those tanneries built within the past ten years, there has been no consistent and harmonious plan of construction. With a fuel so abundant, and withal so worthless, economy in the consumption was best studied by practicing the most wasteful methods, and hence it has been customary for tanners to economize in the construction of their boilers at the expense of their fuel. "Log" or "cylinder" boilers being much cheaper and more durable, they have in many instances been introduced instead of "flue" or "tubular" boilers, not because they were more economical, but in truth because they were more wasteful of fuel. Several of our tanners have been compelled to construct brick furnaces outside and independent of their ordinary boilers, in order to "get rid of" their surplus spent tan. As time goes on, however, and as we come more generally into the use of mechanical power, there will be less solicitude to get rid of this refuse without bringing it within our service. No town or city is without its mechanical appliances, and if the power or heat is not required for the tannery, it can be applied to some other purpose. It may be sent off to the distance of several hundred and even several thousand feet to drive an elevator or an engine, or warm some dwelling or manufactory. There is no reason why all the power should not be utilized, and hence the writer would strongly commend the use of tubular or at least flue boilers in all cases.

As the subject of burning wet spent tan, both in its economic and scientific relations to the tanner, is considered in a subsequent portion of this book, the subject may be dismissed for the present with the statement, which seems necessary in this connection, that the spent tan can be so utilized as to give all the power required by the tanner in softening his dry hides, in grinding his bark, in rolling and finishing his leather, in warming his building, in heating his liquors, and in doing such other service as may be done by mechanical power.

The glue stock, of which every tanner has more or less, is valuable almost in proportion to the care bestowed upon its preservation, and this is true without considering the higher grades of gelatine which are made from calves' pates and feet, and the sinews from neats animals. These are, or should be, preserved by the slaughterer, and handed over to the manufacturer of gelatines for human food. In the preservation of tanners' hide offal, as it is known to come from ordinary dry and green hides and skins, it is of the first importance that the "glue" or "size" pieces should be sweet and pure from all "taint" or "smell." The paper makers use animal "size" very extensively, and such as they use must be absolutely free from all bad odors, otherwise the paper is rendered worthless. With all the care which the paper makers exercise they do occasionally use "tainted" size, and the odor will cling to the paper forever, greatly lessening if not entirely destroying its value. Next in importance to paper size comes the fine glues which are used in book and furniture work. In these, too, it is of great importance that there should be perfect purity and freedom from all smell. It is only for the most common use, such as preparing petroleum barrels, sizing woodwork, etc., that glue that has been in the least tainted can be used; therefore it is that the value of this stock depends almost entirely upon the care taken in its preservation from all bad odors.

When the hides or skins are either fresh from the animal or have been properly cured with salt there is no difficulty in avoiding all decomposition—indeed, nothing but absolute carelessness can bring harm to the "glue pieces." But if the pelts have been dried, and require to be softened before being limed or sweat, then it is that care and close and prompt attention should be given to the preservation of this valuable offal. Whether from green or dry stock, the trimmings should be thrown into a weak lime as soon as they leave the beam, and should be retained in this lime until the hair will almost drop off; when in this condition they should be thrown into a revolving wheel, or may be put in the hide mill, and worked until all the hair has been separated from the pieces. If a bountiful supply of water has been allowed to run on while the wheel or mill was in operation, all the hair will have been worked off and separated from the pieces, which will then have parted with so much of their lime as to make them, when dry, flinty and hard; to avoid this they should be thrown back into the lime for a few days, and again "raised." Before they are taken out to dry they should be thoroughly washed. The drying should be in the open air, and if on a flat board surface then the pieces should be frequently turned. Care should be taken to wash off all the loose lime, so that the pieces may present an attractive, uniform white clean surface. When they are fully dried they should be pressed into uniform bales. Under no circumstances should any tainted or damaged piece be allowed to go in the bale. Paper makers will use such hide offal for sizing, and pay three or four cents per pound more for it than glue makers can afford to pay. Calfskin shanks and pates are worth more for this purpose than hide cuttings, and should always be kept separate.

Until within a few years cattle and calves' hair have only been used for plastering purposes, and for this the harshness produced by the liming process rather added to than detracted from its value; but more recently this hair, particularly the calves' hair, has been used for cloth, carpets and felting, and for these purposes it is required to be soft, which necessitates freedom from lime. Probably it will be ultimately demonstrated that the hair from hides that are unhaired by sweating, so far as the length (as in winter and spring hides) will serve, is far superior to limed hair for this very reason. Millions of pounds of this valuable product, that should have at least gone on the "fertilizing heap," if its length had been too short for cloth or felting, have been washed down the streams.

It is important to select the white hair from all other colors, as it is worth nearly or quite double the price of the general stock. This selection is not attended with much inconvenience if done while the beam work is in progress. If the hide or skin is white, or mainly so, it should be thrown to the beam hand whose duty it is to unhair this color; when he has removed all the white hair he passes the skin or hide to his companion, and so in turn they pass to him all their pelts that have any considerable amount of white hair. In this way a single beam hand accumulates at his beam all the white hair, and of course it can be easily dried and kept separate.

The cleansing of the hair from the lime is attended with little or no expense. Throughout Europe it is washed by hand, large splint baskets being used; filling a basket about half full, an attendant, standing on the bank of a stream, plunges it in and lifts it up until the currents of water passing through carry off most of the lime. But in this country a vat with a false bottom is used to better effect. The vat is placed so as to command a bountiful supply of water, and it has a false bottom bored full of small holes—large enough to allow the lime water and small particles of lime to pass through, while retaining the hair. While the hair is covered with water, and is being plunged, the openings in the bottom must be closed; when the agitation of the contents is completed, the plugs may be drawn from the bottom, and, practically, the particles of lime having gone down below the false bottom, the hair will mat together, settle down and rest on the false bottom; very little or none of the serviceable long hair will have passed out. This process of filling, plunging and draining off may be repeated a dozen times during the day, and the hair will be fairly cleansed. If it is desirable to cleanse the *fiber* of the hair from the effects of the lime, warm water run on the last time, and allowed to remain for a short period, will greatly aid this result. This latter process will leave the hair much softer than if cleansed only with cold water. When it shall be established, as now seems probable, that all calves' and cattle hair can be utilized in making coarse blankets, cheap carpets and felting, other methods for neutralizing the lime will come into vogue.

Never until within the past few years has the hair from hides unhaired by sweating been saved. Whether the hair was long or short it has all been allowed to pass down the stream from the mills, or if saved from the beam it has, in exceptional cases, been used as a fertilizer. It is now demonstrated that hair from such stock is more valuable for all the purposes above referred to than when limed, and when the long haired hides are milled by themselves the hair will always pay the labor of saving. For felting, all hair from hides unhaired by sweating, both long and short, is serviceable, and should never be wasted.

The principal expense in saving hair heretofore has been in the drying. This is obviated by a very simple contrivance, as follows: Make a plank box eight or ten feet long, four to six feet wide, and say two feet high; under a false bottom run a net work of steam pipe; place the pipe midway between the bottom and the latticed false bottom, so far distant from each that a broom may freely pass to clear out the accumulating lime dust—and when thus prepared turn on the exhaust steam from your nearest engine, allowing the steam to condense through this piping and under this false bottom.

Hair may be thrown into this box to the depth of one foot, and will only require occasional stirring for an hour to make it thoroughly dry. In this way the hair may be dried ready for packing, while occupying but a small room space and at an inconsiderable expenditure of labor.

Hair may be packed for transportation in bales, pressed as hay is (indeed, hay packing machines may be used for the purpose), or in sacks made of burlaps, as wool is usually packed. If pressed in bales, refuse boards should be placed on the bottom and top, and the whole bound with wire, precisely as hay is now brought to market. When thus packed, fully 500 pounds can be placed in a bale four feet square.

The water from the soaks, as well as the exhausted lime and bate liquor (including the wash from the hide mills and beam house), should all be collected together in a capacious reservoir situated below the level of the tannery foundation, so that all these valuable liquid manures may be brought together without the labor of elevating. This reservoir may be at any distance from the beam house; indeed, it is far better to be removed a few hundred feet than close by. It should be capacious enough to hold not only the liquid refuse, but also all solids which come from the sweepings and scrapings of the beam-house floor and drying lofts. So valuable are these washings that the writer has known a small tanner to fertilize a hundred acres by the refuse of a five thousand hide tannery. Old spent tan, fine chips or shavings, and even earth, may be carted and thrown into this fertilizer reservoir, and allowed to absorb the ammonia, and then removed to the land, with great profit.

A tanner with poor land about him, who allows his refuse liquids to run into the stream, is culpably thoughtless; and yet such instances are numerous, notably so in the State of Maryland, where the writer has seen hundreds of loads of the residuum of spent limes and bates piled up around the tannery. As the law forbade the tanner from throwing these objectionable substances into the stream, and as he owned no land, and was surrounded with shiftless farmers, large piles of the most valuable fertilizers remained to obstruct his way. The single tannery to which allusion is made could fertilize, if all liquid and solid matter of the beam house was distributed, more than 400 acres of land.

CHAPTER XVIII.

TANNING MATERIALS.

DIFFERENT KINDS OF HEMLOCK BARK—INFLUENCE OF SOIL AND CLIMATE ON THE QUALITY—HEAVY AND LIGHT BARK—VARIETIES OF OAK BARK—THE "SECOND GROWTH" BETTER THAN THE FIRST—GAMBIER—ITS GROWTH AND PREPARATION FOR MARKET—ITS COST COMPARED WITH THAT OF BARK—VALONIA, DIVI DIVI, MYRABOLAMS—"SWEET FERN," ETC.

On the American Continent, bark is the principal material for sole leather tanning, and although many attempts have been made to supersede its use, they have for the most part failed. The reasons for these failures are not very difficult to find, and should convince that large class of experimenters who rely upon "sweet fern," "salt," "alum," "terra japonica" and the like substitutes that the day is far distant when, in this country, any other substance than bark tannin will be successfully used for manufacturing leather.

The tannin from hemlock bark, probably, tans eight-tenths of all the leather made in America, and notwithstanding the opinion entertained in England—and in this country too, by a few—that hemlock is inferior to oak, as a tanning material, the writer believes that, so far as the question is susceptible of demonstration, this opinion is erroneous.

But to pass, for the present, from the consideration of the comparative merits of the manufactured article, and consider only the comparative cost and value of the bark as a tanning material, it should first be stated that there are two kinds of hemlock—*white* and *red*—and although there is a great difference in the timber of the two kinds, there is not much distinction in the tanning qualities of the bark. Some tanners think there is a dissimilarity in strength, and also in color, but the conceded difference between hemlock barks may be attributed to other causes than this difference of "species."

Soil and climate we know influence the tanning qualities of all barks, and none more than the hemlock. All tanners know that hemlock bark from low, marshy ground, or the swales of the hills and mountains, is much thicker than from the high lands and mountains, and further, that a southern exposure produces more thrifty trees, and consequently thicker bark. The wash of the mountains, of course, feeds and enriches the valleys, giving richer soil and damper earth, both of which are indispensible to rank vegetable growth. It may not be generally known that all barks, like all woods, are made up of distinct layers or deposits, one layer being formed each year. This formation takes its character from the soil and the season; if in rich soil and damp, wet surroundings, the layers will be thick, and "fat" with tannin, but if otherwise, the layers will be thin, and the tannin "lean." So uniform and unerring are the laws governing these deposits, that naturalists claim they can not only tell the exact age of the tree, but they can determine with considerable accuracy the nature of the seasons through which the life has progressed. Assuming then that tannin is deposited just in proportion as the growth of the bark or tree is rank and healthy, we can easily see how it happens that in some sections of the country a cord of hemlock bark will weigh 1,600 to 1,800 pounds, and tan but 140 to 150 pounds of leather, while in other sections a cord will weigh 2,200 to 2,300 pounds, and tan 200 pounds of leather. These may be taken as extremes of quality, but to find these extremes one need not go out of the same neighborhood. The swamp bark represents the one extreme, and the high mountain bark, particularly if having a northern exposure, the other extreme; and yet, while this difference is known to all well-informed tanners, they continue to pay the same price for each kind, and wonder why one cannot tan as cheap as another. But there is a further consideration which weighs against the light bark, and it is this: the small, stunted growth bark will *curl*, forming "gun barrels," while that from the heavy swamp timber will lie flat. This fact goes, to some extent, to make up the great difference between the cord of 1,600 pounds and that of 2,300 pounds, but the assumption of inferiority is outside and independent of this disparity of measure; if the premises are right, 2,000 pounds of heavy hemlock bark will tan more than 2,000 pounds of light hemlock. Practically the difference cannot be very great, but so long as tanners continue the unjust practice of buying bark by the cord, instead of the ton, they will do themselves this wrong. The writer knows of a location where, for many years, the competition for bark has been very strong; but one tanner, appreciating the difference above referred to, has always selected and purchased the lots of heavy bark, paying from one to four shillings per cord more than his neighbors, by which means he has, probably, paid less for the actual tanning material than any of his competitors. Why should one bark peeler be required to rank and draw 2,300 pounds from six to ten miles for the same price that another gets for 1,600 pounds? The injustice both to the tanner and bark peeler must be very apparent. The light hemlock bark is supposed to impart less coloring matter than the heavy. This is possibly true; indeed, it is so in the judgment of some of our best-informed Eastern tanners. Acting on this supposition, the fair leather tanners of Connecticut and Massachusetts continue to use the light hemlock obtained from the upper counties of Massachusetts and Vermont, supposing that the want of coloring matter will compensate them for the deficiency of strength. The beautiful lemon color produced by the Connecticut and Massachusetts sheepskin tanners may be the result of the *weak* but *pure* liquors which they use, but if they subjected their bark to the same *heating* and *leaching* process as that of our sole leather tanners, possibly the difference would not be as great as is now imagined. However this may be, the supposed difference in the coloring matter of hemlock bark is keeping several large fair leather tanners in Connecticut from availing themselves of the cheaper hemlock of Pennsylvania and New York.

To verify the theory here presented in regard to the quality of bark, let the reflecting tanner pass over in his mind, as he readily may, the whole tanning region of the country, commencing with Maine, on the East. There we find a cold climate, and for the most part a poor soil. The valleys of the Aroostook, the low lands surrounding Moosehead and other lakes, may be exceptions; but generally these rich lands are covered with pine, and have only occasional patches of hemlock. The hemlock bark from that State is thin, and will not weigh, as bark is ordinarily measured at our tanneries, over 1,800 pounds to the cord. The same general character holds true of the bark of all the New England States. In the lower counties of the State of New York we have a warmer climate and a more diversified soil; where the soil is favorable we have thick bark, but the hills and mountains give us much thin bark. This is true of Northern and Eastern Pennsylvania also; but as we proceed West and South in New York and Pennsylvania, the country is more open and level, the soil richer, and we have a more uniform and much thicker bark. Take those counties where wheat grows naturally, not merely on the river bottoms, but on the slopes of the highest hills and mountains—where you can see no "barren wastes," as in Eastern and Central New York—and you are sure to find thrifty trees, with long bodies and thick bark—bark that will weigh 2,200 to 2,300 pounds to the cord.

The effect of soil and climate on the growth and strength is still more apparent in oak bark. The bark from this tree is comparatively worthless in a cold latitude, while in Lower Pennsylvania, Maryland and Virginia, it is strong. There it grows on the bottom lands; this, with the warm climate, gives the rank growth. But in New York the oak grows only on the hills and ridges, being crowded out from the swales by the hemlock, birch, beech, and maple, so that we cannot say what it might do under more favorable circumstances.

It was suggested by the late Col. Pratt, and has been reremarked by other observers, that sea air has something to do with the strength of bark. This observation, confined to oak, certainly seems to be verified by the known facts, for the oak of Ohio, Michigan and the Western States generally is very weak—scarcely strong enough, as now used, to furnish liquors of sufficient strength to preserve the hide from putrefaction. But may it not be that the oak of those States is of a different kind from ours? The writer has not seen rock oak there, but has seen white oak and red oak of immense size and growth. This white oak bark is a poor tanning material, and the red oak but little better—scarcely yielding enough tannin to pay for peeling. May it not be that the influence of sea air makes the difference between their oak and ours?

How many pounds of leather will a cord of hemlock bark make? This question, a thousand times asked, has never been answered, and never can be, until a more certain rule of measurement is determined upon. As already hinted, if in a new location, where strong competition does not compel the tanner to take 90 to 100 feet for a cord, but where he can get 128 solid square feet, and that, too, of good, heavy bark (and the heavy bark is always peeled first, for obvious reasons), then 200 pounds of leather can be made from one cord of bark; but if these conditions are otherwise, then from 130 to 150 pounds—and it is safe to say that throughout the old tanning districts of New York State 160 pounds is a full average.

Little need be said of oak bark, for its merits are appreciated the world over. In this country we have the following kinds: *Rock Oak*, (called by some "chestnut oak;") *Yellow Oak* (called sometimes "black oak," from the dark, black exterior, but by tanners called "yellow," from the fact that a very yellow color is produced by its bark;) *Red Oak* and *White Oak*. The first two are considered the most desirable. Their strength is about equal. The coloring matter, however, of the yellow oak is so objectionable for sole leather that, unless used with red oak or hemlock, its value is very much impaired, but it is valuable for dyeing purposes, and is exported from Philadelphia and Baltimore in large quantities, under the name of "quercitron bark." This bark (although of rather inferior quality) is found in large quantities in the State of New Jersey.

White Oak makes a beautiful leather, but the tannin it contains is so small in amount as to render it almost worthless—certainly not more than half as valuable as the better descriptions of oak. The coloring matter, too, although very good of itself, is too weak to be of much value as a dye.

Red Oak is a heavy, hard bark to manage—difficult to grind, and heavy to handle—and when leached or used as a "duster," produces a "mean" red color, without the "fresh hue" of the hemlock, or the "true bloom" of the oak, and would seldom be used for its own sake—but the wood is valuable for staves, and hence the bark is peeled, and finds its way to the tanner, and, to his detriment, he is often induced to use it.

The original oak—that is, the first growth, even of the best descriptions, is quite inferior as a tanning material; but the second growth—the young trees, say from fifteen to twenty-five years old, particularly of the rock oak—is very valuable, and, when properly blended with hemlock, makes a leather which, for color and wear, will compare favorably with the best tannages of the world.

The effect of age upon the oak tree is to cause it to throw off a *dead ross*, which loses its tanning qualities; hence, until within a few years, it was customary with all our oak tanners to ross their bark—throwing away as worthless fully one-third the substance of the bark; but, after using up the old first growth, and coming down, as we now have, to the second growth, which has much less of this dead ross, the practice has gone into disuse.

Oak trees reproduce themselves in about twenty to thirty years, on good soil; hence we may safely rely on a continuance of a supply. This circumstance, coupled with the favor with which mixed bark tannages are regarded, leaves no doubt that for all time we shall have a full supply of oak bark. Indeed, it has been estimated that more oak tannin is now on the trees of Ulster and Greene counties than there was twenty years ago. Although this reproducing quality does not belong to hemlock trees, yet, in such abundance are they, both in New York and Pennsylvania, that many hundred years must pass before they become extinct.

The curing of oak bark is an important subject, and is but little understood—or, if understood, is but poorly attended.

to. Fully one-third of the strength is lost in the careless cure, exposure to wet weather, piling damp, etc. The English pay as much attention to the cure of their bark as to that of their hay. The damage to oak bark, from improper exposure, is quite as ruinous as to hay, and yet our bark peelers give the bark only as much attention as their leisure will permit.

The substitutes for hemlock and oak bark—the two principal agents used for tanning in this country—are very numerous.

Gambier, or terra japonica, stands first and most prominent among the substitutes for bark, both in England and here. In 1840 the first "terra" was used in America, and only in small quantities, a few bales then sufficing to supply the market; now the import amounts to thousands of tons annually, and its use extends to all the northern Atlantic cities, Newark, N. J., and Danvers, Mass., being the largest consumers.

By a decision of the United States Treasury Department, gambier, catechu, cutch, and terra japonica, are all considered in commerce as essentially the same, but while they all come under the common name of terra japonica or gambier, a preparation of catechu, there are three different kinds of gambier, (in bales, cubes pressed, and cubes free), all of which differ in price from each other, and from what is known as cutch. The supply comes almost entirely from Singapore, in the East Indies, that island having become an entrepot in which are collected for exportation the productions of Cochin China, Siam, the Malayan peninsula, and the whole region of the Eastern Archipelago, from Sumatra to the meridian of New Guinea and the Phillipines.

The product, as we receive it, is extracted from the leaves of a species of acacia, those from different parts of India giving slight variations in the quality. The shrub, or bushy tree, is grown from slips or cuttings, on plantations. It is cultivated and manufactured on land leased from the various Rajahs. The shrub is allowed to grow for the term of two or three years, till it attains the size of dwarf pear trees, when they commence to strip the leaves, stripping at all seasons of the year thereafter, for five or six years, until the soil is exhausted; the leaves come thickly on the long, drooping limbs, which grow outward and downward from the ground to the top. The leaf is in shape like the mulberry, but is thick and gummy to the touch. The production can be increased to any extent, and would, probably, if a long ruling of high prices should warrant it. The cost—the ground rent being nominal—is the labor, fuel and transportation, at the plantation not over 1c. to 1½c. per pound.

The leaves are boiled to extract the gambier; when reduced to the thickness of molasses the leaves are raked out, being used for a land dressing about the pepper trees; the water is then evaporated from the gambier by placing it in pans and exposing them to the heat of the sun; later, after drying the gambier in these pans, it is cut into squares, forming cubes, which, when dry enough to handle, are bagged, sent to market (Singapore), screwed into bales of 266 pounds, matted, covered with gunny, and is then ready for shipment.

In the setting out of the cuttings they are placed as near each other as will leave room for the shrub to grow. An acre will produce 2,600 pounds annually, but it soon runs out the soil; years ago it was mostly produced on the island of Singapore, but now very little is raised there. There are several kinds or qualities, the Rhio being the best; it originally came in baskets, in loose cubes, but lately, like the others, comes to market in bales.

Cutch and gambier are from the same or similar trees, but the two do not come from the same part of the country.

Cutch is extracted from wood of the larger and older trees; these trees are not cultivated, but taken naturally, the heart cut out, and from that the article known as cutch is extracted.

But to pass from the consideration of the source and manufacture of gambier, let us consider it as a substitute for bark. In its pure state it is largely used for dyeing purposes. The color is so dark and unsightly that tanners cannot use it—unless with yellow oak—without leaving a very objectionable color. The strength of pure gambier has never been appreciated in this country, and, probably from the consideration above presented In England, on the contrary, where oak bark is used with a "strong bloom," the color is overcome.

Let me say here to the tanners of our country that gambier, in all its varieties, is an expensive substitute for bark. Hemlock bark at $10 per cord is as cheap as gambier at 3½ cents per pound; and yet the Newark tanners, with bark at $9 to $12, are using gambier extensively, with the latter costing from 5 to 7 cents a pound. The same may be said of the tanners of Danvers. Why this want of economy? Because gambier comes in a form very convenient for use—can usually be bought in any desired quantity, at any season of the year. Its use prevents overcompetition for bark, and as the tanners are not *quite sure* that it costs them more than bark, they keep on using it, on the principle they do much of their business, viz.: "All is well that ends well." There is a day in the future when all this blind waste must give place to true economy; and when that day shall arrive, perhaps to be hastened by the substitution of extracts from the barks of our own forests, then shall competition from England be no longer feared by us, but free trade in leather, as in everything else, shall best serve our interests, as it would certainly best comport with our self-respect.

The exports of gambier from Singapore for the past sixteen years have averaged, for each year, as follows:

To Great Britain.	15,818 tons.
To the Continent	3,930 tons.
To the United States.	4,818 tons.

According to these figures the amount taken by the United States has shown very little variation during the whole period. The amount used by Great Britain is supposed to be equivalent to about 60,000 cords of bark yearly. The price in London is now from £25 to £32 per ton.

In Great Britain they have many other tanning agents for their heavy leathers, with which in this country we have so little experience that it would be presumptuous for the writer to speak particularly of them. They are mainly valonia, divi divi, myrabolams, etc. When the English tanner shall become as communicative as he is at present reticent, the whole world will be enlightened as to the comparative economy of these agents with those in use with the tanners of the rest of the world.

Valonia is the commercial name for the acorn cups of an Asiatic species of oak, which forms a very considerable article of export from the Morea and the Levant. The cup only forms the valonia, the acorn not being exported. While kept dry it presents a bright drab color; exposure to dampness makes it black and destroys its tanning properties. It is very light and bulky, making the cost of its freightage high. It is very little used in the United States, but in England the imports are about 4,000 tons annually. The price in London is now from £15 to £18 per ton.

Divi divi is a pod of a shrub, a native of South America and the West India Islands, the tannin of which is concentrated in the rind of the pod, immediately beneath the epidermis; the inner portion, including the rind, is worthless for tanning. The leather prepared with divi divi is likely to be porous, and tinged with brown, or brownish red; but little of it is used in this country.

Myrabolums is the commercial name of the dried fruit of the *meluccanna*, imported mainly from the East Indies. The imports at London for 1875 were 9,800 tons, against 11,200 tons in 1874, and 4,100 tons in 1873. The present price is from £13 to £17 per ton.

In addition to the above tanning materials—so largely used in England, and in combination one with another as well as with bark— he English tanners also use considerable quantities of Mimosa, Belgium and Cork tree bark, and are yearly taking constantly-increasing quantities of our hemlock extracts.

Sweet fern is a tanning shrub or plant found on the barrens of most of the counties in the Eastern States. It has been extensively used in England, and to a limited extent in this country. The writer has been reminded by frequent circulars of various patents granted for the use of this plant for tanning. After reading the strong array of certificates in favor of the excellent quality of leather made from "sweet fern," by postmasters and other equally good judges, he does not dare to question the value of such patents. There is one point, however, to which the attention of all experimenters on these substitutes should be called. The question is not whether you can make tough leather with "sweet fern," sumac and the like, but whether you can make the same or better weight. Can you make the leather with tannin so that it will both resist *friction* and *water?* And more important than all, will the cost be less than with hemlock bark at $7 per cord, or good rock oak bark at $8 per cord or ton?

That leather can be made from sweet fern, sumac, birch, chestnut, willow—indeed, almost all barks—and that various acids and salts will *cure*, or, if you please, *tan* leather, no one at all acquainted with the subject will dispute. Some of the toughest, best working calfskins that were ever produced in our market were tanned with birch bark, and the color, too, was good; but they were tough because the bark liquor was weak and the fiber, in consequence, elongated; the weight and general plumpness were sacrificed to toughness. But when shoemakers, or even postmasters, certify that calfskins resist water better when tanned by these processes than when tanned by pure oak or hemlock bark, the writer is willing to believe them honest, but attributes a little of their zeal to kindness of heart rather than maturity of judgment. This subject is well illustrated by repeating a conversation held not long since with an old gentleman, who prided himself on knowing "a little about leather," as on other subjects he was wise. He said he always bought his calfskins of the "Shakers," for, said he, "they tan without steam," and "when my boots are made of their lightest calf, I can wade all winter through snow and water, and do not have even damp feet." The old gentleman meant to tell the truth, but probably he had not been in the snow and water without rubbers for years. Yet he succeeded in his purpose—he entered his protest against "steam tanning." It is thus that many people, from only a partial understanding of a subject, are free to give their opinions, upon which opinions too many, equally credulous, confide, to their cost.

Many old-fashioned tanners, who have used only weak bark liquor, and for the first time tried terra, are astonished at the result. They tan as much in sixty days as by their old system they did in six months. The solution of the matter is plain; they make a strong decoction of the terra, one that will stand 20 to 25 degrees by the barkometer; whereas, by the old bark process, they were trying to tan with 6 or 8 degree liquors. But let any tanner take one hundred pounds of terra and dissolve in the usual way, and then take the strength of one cord of good oak or hemlock bark, and he will find the latter will tan double the quantity of leather that the former will.

In this chapter the tanning agents employed in tawing or tanning light leathers have not been considered. The omission has been made designedly, for the tanning and tawing of light leathers is a trade by itself, the treatment of which is not within the scope of the present treatise.

CHAPTER XIX.

THE COST OF TANNING.

THE SEVERAL ITEMS VARYING WITH DIFFERENT TANNERS—DIFFERENCES FROM UNEQUAL WEIGHT OF THE CORD OF BARK—THE AMOUNT OF TANNIN IN UPPER LEATHER AS COMPARED WITH THAT IN SOLE LEATHER — COMPARATIVE COST IN MAKING HEAVY AND LIGHT GAINS — THE THEORETICAL STRENGTH OF BARK NEVER REALIZED—COST OF "UNION" AND OAK TANNING—ESTIMATED COST OF TANNING IN EUROPE.

Approximately, the cost of tanning is as follows:

Hemlock sweat sole leather.......... ....℔ lb	6@ 7c.
Union lime sole leather	8@ 9c.
Oak lime sole leather.....................	9@10c.
Oak lime rough leather....................	8@ 9c.
Hemlock lime rough leather...............	6@ 7c.

The varying circumstances under which the tanners' profession is pursued will cause the cost to differ within the limits above indicated. When a closer estimate is desired, giving the cost of each department, then a great diversity of opinion prevails. One tanner devotes extra time to the beam-house work; another to the finishing; still another to the handlers or layaways. If bark is cheap at one location, inland freight and cartage is in excess of that at another place where bark is dearer; therefore, in stating the elements of cost, it is understood that an average is struck. These attempted details are of far less value than the general conclusions, which may be relied upon as above stated.

One ton (2,240 pounds) of average hemlock bark will tan 200 pounds of sole leather. Some of the exceptions are as follows: 1st. Where the bark is ground and leached imperfectly, or in an extraordinarily perfect manner. 2d. If the leather is tanned with very strong decoctions, and thereby a very large gain is obtained, as against weak liquors and a light gain. 3d. Great delays and wastes in applying the tannin to the leather—delays which induce the formation of gallic acid, or the bringing of the fresh, sweet, strong decoctions into contact with liquor which has already formed a large proportion of acid.

These exceptions cannot be always anticipated or known; it is safe to say, however, that they vary the result all the way from 180 to 200 pounds of leather made from one cord or ton of bark. If tanners buy their bark by the cord, and get less than 128 feet solid measure—resulting, as often happens, in getting 1,800 pounds, instead of 2,240 pounds for a cord—then, of course, such tanners will find the results of their tanning to vary still more.

Among the topics above suggested there is but one to which attention will be now specially called, since on that one depends, far more than is generally supposed, the profit or loss of the tanner, *i. e.*, the strength of the liquor employed.

The upper leather tanners of New England, who pay from $10 to $12 per cord for their bark, claim to tan and probably do tan from 300 to 400 pounds of upper leather with 2,240 pounds of bark. If we comprehend how this is possible, it will enlighten us as to the point under discussion. The upper leather tanners draw their tanning and coloring matter from agents which furnish about 20 per cent. extractive material, while the vigorous sole leather tanner obtains his capital mostly from the 7 to 8 per cent., tannin which the bark contains. The tannin gives all the gain added to the gelatine, but the coloring matter permeates the fiber, while cumulative gallic acid holds it from decomposition. Upper leather, then, is not tanned and filled as sole leather is, and to this extent, and for this reason, bark extract will spread itself over far more fiber when all the extractive matter is employed, than when it is so manufactured as to hold only the tannin pure and simple.

The calfskin and upper leather tanners of Germany, Austria and Switzerland make fully 400 pounds of rough stock from a ton of the best coppice oak bark, and where they use the "larch," corresponding to our "spruce" bark, they probably make about 200 pounds (this bark having less than half the strength of the former).

If upper leather and calfskins are to be sold by the pound (waiving the question of quality, especially toughness), then it is evident that these light tannages cannot be afforded. But if sold by measure, then a light tannage is profitable for both the tanner and the consumer, under proper circumstances. It is not proposed here to discuss the question whether a light tannage will resist water. A tannage with an elongated and merely colored fiber will not carry stuffing when curried into upper, and will not resist dampness, and for these reasons such leather should not be used for the common wear of the people, whose feet are exposed to the varying conditions of the weather in our moist and wet climate.

This preliminary discussion has been introduced to solve the question as to whether a sole leather tanner who tans by the pound, and is not interested in the question of interest, can afford to make heavy gains for his employer. One tanner makes 160 pounds and another 175 pounds of leather from 100 pounds of the same description of hides; are they entitled to the same pay per pound? Is the cost to each proportionate? The argument on the one side is as follows: It costs a certain sum to work in, handle and finish a given lot leather, whether of a heavy or a light tannage; the cost of the bark being alone considered, it cannot exceed and most usually falls short of the price received for tanning, even though that price is as low as six cents per pound. The sole leather problem is, then, in fact, but the upper leather question over again, which would ask and determine the following: Can a tanner afford to tan rough leather for less per pound than sole leather, less the finishing? It is no answer to say they do tan it for less, and it is not convincing that small yards in old tanning districts, without much interest to pay, are still pursuing this trade successfully. The argument on the other side is that heavy gains cannot be made without strong liquors; strong decoctions cannot be obtained unless more or less waste is permitted—waste in the liquor itself, and more strength lost in leaching. Besides, the actual net added weight costs more than is received, which is calculated as follows:

Original weight of hides.............	100 pounds.
Less hair, grease, flesh, etc...........	15 pounds.
Net gelatine and animal fiber.......	85 pounds.

Now, whether this product is raised to 160 or 175 pounds is a question of mere intrinsic cost of the pure tannin which is *capable of combining with the gelatine*. In the one case 75 pounds is required, and in the other 90 pounds. The cost of these respective factors made from bark at $6 per cord would be (on the theory on which we are proceeding) fully 8 cents per pound, since in both cases we start with the hide capital of 85 pounds. It may be assumed that there is a discrepancy between the theoretical and practical percentage of tannin obtainable from bark. We know that 2,240 pounds of hemlock bark will make only 200 pounds of leather. Chemists tell us, however, that there is 8 per cent. tannin in this bark; consequently there is in this ton of bark 156 80-100 pounds, which, combined with 85 pounds of gelatine, should give 241 80-100 pounds of leather. What has become of this 71 pounds of lost tannin? When any tanner will practically solve this doubt, then, and not until then, will the main question be answered. Practically, the cost of a pound of tannin is from 6 to 8 cents, and not 3 to 4 cents, as is generally supposed.

There is another subject nearly allied to the cost of bark, and that is the intrinsic cost of the oil put on sole leather. If the oil added gave full weight, then it could easily be determined whether it was profitable for the tanner to put on much or little oil. But, like the tannin just considered, there is much evaporation and loss, and the extent of interest in the final result of weight and profits must determine whether the tanner can afford to put oil on his sole leather at all, and to what extent.

The cost of administration in a tanning establishment is always underestimated, and from this source, in the writer's judgment, many of the discrepancies result. A story is told of a young man who left his father's home to reside in the city of New York, promising to keep an accurate account of his expenses, and, with all his care, he found himself overdrawn more than $100 at the end of the year. To the inquiry from his father as to the cause of this discrepancy, he said it must be "litterateur." The incidental expenses of a tanning establishment are not less difficult to define. If an attempt were made to properly apportion each item of expense in the production of leather, the factors of the total cost would be about as follows:

Cost of bark (hemlock) per lb	3c.
Cost of soaking, milling, sweating and beam work	1c.
Yard work, including handling, laying away, etc.	½c.
Finishing, including drying, rolling, etc	½c.
Insurance, interest on tanning and bark	½c.
Freights to and from the market	1c.
Administration	½c.
Total	7c.

The cost of union or pure oak tanning will vary considerably from that of hemlock tanning where the hides are unhaired by sweating, for, besides the added cost of bark, the extra care throughout the whole process, including the labor expended in the preparation of the hides, that are mostly green or green salted, will make up the difference, and the cost of the several kinds may be considered as stated at the head of this chapter.

After the most diligent inquiry the writer has found it impossible to even approximate the cost of tanning in the principal nations of Europe. A London tanner in 1873 estimated his tanning to cost 12 cents per pound in bark, and 10 cents per pound in terra japonica and valonia, but confessed that it was only an estimate. More recently a Bristol tanner calculated the cost of his pure bark tanning at 15 cents per pound, and even this is probably only an estimate. The cost of coppice* oak bark in the most favored locations is not less than $25, gold; and in most sections of Europe it runs up to $30, and even $40, per ton. On light tannages no doubt this bark will go a great way, but on butts and bends it may be assumed that foreign tanners use liquors

* "Coppice" bark is from small trees, too small for timber, say six to eight inches in diameter; also from limbs, equal to our "second growth" bark.

of the highest strength. This is particularly true of the tanners of Great Britain; perhaps it is not true to the same extent of tanners on the Continent of Europe.

The cost of all other material, including labor, is not greatly different in Europe from the cost here. The tanners of Great Britain pay about $1 per day for their average hands, while in other countries they pay, according to a recent authority, about 80 cents. But it may be fairly questioned whether our labor is not more effective, especially in view of the increased amount of labor-saving machinery which we employ over some, if not all of our competitors.

CHAPTER XX.

QUICK TANNING PROCESSES.

COMMON ERRORS OF THOSE OUTSIDE OF THE TRADE—HOW WORTHLESS PATENTS ARE MULTIPLIED—EXPERIMENT IN TANNING BY HYDROSTATIC PRESSURE—VACUUM TANNING—DIFFICULTIES ATTENDING THIS METHOD—HOW AGITATION OF THE FIBER FACILITATES THE PROCESS—A GENTLE MOVEMENT, WITH OCCASIONAL REST, MOST EFFICACIOUS—TANNING VS. TAWING.

It is noticeable that most of the attempts to substitute new for old methods of tanning are made by men outside of the trade. Very few of the new inventions for tanning, particularly those that contemplate the saving of time, originate within the trade itself. The inspiration of all this solicitude on the part of the outside world seems to come from the idea that tanning is a slow and tedious process, which needs invigoration by the genius of inventors and men of thought. They have "read history," in which it is claimed that seven years is the allotted time to tan butts and bends and make good leather, and that in this country we have only improved on this time by the introduction of "steam" and other "forcing expedients," which render the leather products here much less valuable than in Great Britain.

There is no topic which requires more vigorous treatment than this, not only for the good of the trade, but for the benefit of that large and unfortunately increasing class of men who desire to get a living by their wits and without labor.

As it is now our Patent Office gives encouragement to this class of parasites, for there seems to be no claim for improvement too absurd to receive favor, and patents are multiplied to such an extent that no man can keep an account of them. There are at present more than twenty patents for unhairing hides with alkalies, when the process of taking off the wool from the sheep and the hair from the deerskin by hard wood ashes is older than our civilization. What is most urgently needed in our Patent Office is at least one Examiner that has practical knowledge of the tanner's art. Such a selection it is not unreasonable to ask, when it is considered that this manufacture stands second in importance among the industries of the country. Proper discrimination would greatly aid improvements, while our present system confuses and retards meritorious inventions.

Among attempts to facilitate the tanning of leather, perhaps no method or device has been more seductive than the forcing process known as the "vacuum" method. It is because of its specious character that attention is here given to the details of its history and failure. Hydrostatic power gave birth to the first idea of tanning by pressure. "If," said a student at school, "I can raise myself with one quart of water, by means of a hydrostatic bellows, that principle can be availed of to force tannin into the pores of a green hide," and so the experiment was tried, in the following manner: A keg of very strong construction was procured, and a tin tube one inch in diameter was run up through the lofts of a tannery to the hight of thirty feet; the keg was filled with strong tan liquor, after a green prepared calfskin had first been placed in it, and then the tube was screwed on to the socket and also filled with liquor. The young tanner student had the satisfaction of seeing the liquor forced through the joints of the keg, and finally the keg itself burst, by the hydraulic pressure caused by this small tube. When, however, the keg was burst open, the skin was found to be only colored, and the fiber no more permeated by the tannin than it would have been if the skin had been thrown into an ordinary vat of liquor for the same length of time. This experiment was repeated several times, for a longer period in each case, by applying the power more gradually; but the result each time was the same. Tan could not be forced into the pores of the skin by surrounding it with liquor under heavy hydrostatic pressure.

The next experiment tried by this young student was to take a piece of prepared hide and place it under the exhaust pump. He had seen eggs expand and burst in a receiver, on exhausting the air which surrounded them. He had seen meat and dead animals expand and swell almost to bursting under a similar operation, and so, he reasoned, if he could place a hide in that condition, and produce that effect, and could, at the moment of the expansion, let in tan liquor—the result must be to force the liquor into the most interior cell of the hide, and, as Sir Humphrey Davy had demonstrated that tannin and gelatine, when brought into contact, would both mechanically and chemically unite, he reasoned that such process of exhaustion of the air and swelling of the fiber must result in immediately tanning the hide, on the admission of the tan liquor. But such was not the result. The hide did not swell like other animal substances filled with air. He found that the cells of the hide were filled with water, and that water would not expand in any perceptible degree on account of the exhaustion of the air.

A few years later an English engineer, who had through many years of his professional life practiced the art of preserving wood by kyanizing with appropriate chemical agents, conceived the idea of coming to this country and tanning leather on this vacuum or kyanizing principle. He spent his entire accumulations, amounting to over $10,000, in constructing a large iron tank in the shape of an egg, and lined this vat with copper. This tank was capable of holding one hundred heavy butts, and of resisting a pressure of one hundred pounds to the square inch. He provided himself with very substantial pumps to exhaust from this tank all the air, even to the extent of making almost a perfect vacuum, and other pumps to force in liquor, until he had obtained a pressure of 100 pounds to the square inch. Between these butts he had placed cocoa matting, so that there should be a perfect circulation of liquor. In short, he prepared himself with every facility which money could procure to tan butts in a large and practical way by the exhaust or vacuum process, and after repeated attempts he failed entirely. What induced him to abandon his cherished enterprise was a small incident, which may throw light upon the subject for the benefit of those who yet see, or think they see, a defect in his method. In placing his butts in this egg-shaped vat, he found a space at the top which he could not fill with leather, and he placed there several blocks of seasoned wood, one foot square. When his experiment had ended and failed, after many anxious hours and days of trial, he found the wood perfectly saturated and tanned, but his butts were only colored through the grain. His conclusion was that the force that will kyanize wood will not tan leather.

The name of the engineer was Thomas T. Ferguson, and the place where the experiment was tried was Sparrowbush, Orange County, N. Y. The full details of the experiment were published in the SHOE AND LEATHER REPORTER some years since, it having been made about the year 1855. Mr. Ferguson obtained patents in Great Britain and France, and applied for a patent here, but was so discouraged by this effort as to abandon all further attempts at quick tanning.

After these experiments had been tried and published to the world there came another experimenter, this time with an indorsement of a patent, by which leather was to be tanned in wooden tanks or vats that were far inferior, both in strength and construction, to the plan of the English engineer just mentioned, and many tanners have been induced to invest in the new process. So far as is known, however, the new quick tanning process has been a failure, and it stands so confessed by those most interested.

The *rationale* of this whole matter of tanning by pressure from both surfaces has heretofore been greatly misunderstood. The term "both surfaces" is used, for it is conceded that if force is applied to one side, and that the flesh side, the tan liquor will very readily pass, and tan the fiber with which it comes into contact. This is notably shown in the tanning of morocco and light leathers generally. A goat or sheep skin sewed up and filled with tan liquor will, by the gentle pressure of its own weight, tan in a few hours. A calfskin sewed up and placed where the keg was placed in the student's first experiment will tan in a few moments—that is, tan liquor will be forced through the skin, and the whole fiber will become colored, and even tanned. But it must be acknowledged that this pressure from one side is quite a different thing from equal pressure from both sides at the same moment. Water or fluids of some kind fill the pores of the skin or hide, and these are not compressible, and in this lies the solution of these repeated failures. Water may be expanded by heat, but it cannot be perceptibly compressed by mechanical force or power.

What would be the effect of forcing tan liquor from both surfaces into a hide that had no moisture or water in its fiber? Simply to "tan in" just so far as there was tannin to combine with the gelatine, while the water in the liquor would pass on and fill the inner portion of the hide; and then, before the process could be repeated, this water would be required to be pressed out by mechanical power or dried out by exposure to the atmosphere—and this alternate drying and pressing process would be so tedious and withal so disturbing to the fiber that it would prove impracticable.

It is not claimed here that under the recent vacuum process, so called, quicker time has not been made than by the older methods. The more frequent renewals of the liquors, the greater activity in handling, the agitation of the fiber occasioned by the alternate pumping out of the air and forcing in the liquors—these causes would naturally shorten the process, and are in themselves quite sufficient to account for the more rapid action of the tannin in combining with the gelatine of the hide. But if, notwithstanding these experiments and warnings, tanners will insist on paying their money for patents that are worthless, no one should be held responsible, but our lunatic asylums should be enlarged.

The supplement to Ure's Dictionary of Arts and Manufactures, at page 1044, contains an account of an invention for tanning in vacuo, patented by M. Knoederer, in Bavaria, which should be read by all persons contemplating further efforts to avail themselves of this vacuo principle in tanning. The account, apparently, is simply a statement made by the patentee as to his claim, and the results. There is nothing in either that is worthy of serious consideration, except from the fact that the account has a prominent position in a work of very high authority on general mechanical subjects, and for this reason it attracts attention. It is noticeable that in this process the sides or skins are passed under a mechanical press before they are put into the vacuo vessel. In this manner of preparing the stock, concession is made to the fact that water cannot be compressed.

The writer has not desired to inveigh against the process of tanning by vacuum. If he had wished to show the comparative expense of this method he would have gone into a calculation of the cost of the vats and machinery necessary to carry on such a system, giving the experience of some of those who have tried it—which shows that the vats made of plank and timber, and exposed to the air on their whole internal and external surface, will decay in three years to such an extent as to require renewal—and so far as is known most of those who have tried the process have not renewed their vats after the first set has given out.

One of the most noticeable defects of vacuum tanning is one that has been experienced in tanning by suspension in the ordinary open vats. The leather is "baggy"—that is, the nerve of the hide is tanned in its normal condition, and the shape of the animal from which the hide or skin was taken is approximately preserved. To overcome this defect it has been the custom of those tanners who tanned sole leather by this method to take the sides out of the vacuum vat and lay them away in the ordinary manner for a month or more.

Whether sole leather is tanned by suspension in a vacuum or ordinary vat, the experience is that a much finer offal is produced than results from the ordinary handling and laying away. But the leather is neither so plump nor free from "bag" as if tanned in the usual way.

It is claimed that ordinary slaughter sole leather, weighing eighteen pounds per side, can be struck through in from fifteen to twenty days, if vigorously attended to, under the vacuum process. The leather is then taken out and laid away for thirty or forty days, when it is ready to finish. In making this statement, and conceding all that is here claimed, what is the gain, either in time or quality? In ordinary tanning a daily strengthening and handling will bring this same leather through in sixty days, without the violent agitation and expense attending this forcing process, and with better color and plumper fiber.

The innovations upon old methods of tanning take upon themselves two general forms. They are either physical and outward, or chemical and latent. Of the former, pressure, either with or without the aid of the vacuo principle, has the greatest number of advocates. All, or nearly all, the chemical agents employed in the shortening of the process turn out to be the old methods of tawing rather than tanning.

It is noticeable that even the vacuo principle is greatly aided by *the agitation of the fiber*, for, by reference to the experiments of Knoederer, it will be seen that he tanned in about one-half the time when the leather was agitated in the vacuo vessel, over and above the time taken when there was nothing but the force of the vacuum to hasten the process. The results are reported as follows:

	Time required for tanning in vacuo without motion.	Time required when motion is employed.
Calfskins	from 6 to 11 days.	4 to 7 days.
Horse hides	35 to 40 days.	14 to 18 days.
Light cow	30 to 35 days.	12 to 16 days.
Cow hides, middling	40 to 45 days.	18 to 20 days.
Heavy cow hides	50 to 60 days.	22 to 30 days.
Ox hides, light	50 to 60 days.	20 to 30 days.
Ox hides, heavy	70 to 90 days	35 to 40 days.

Thus it will be perceived that motion or agitation of the fiber is a most essential promoter of quick tanning, even in vacuo. To this fundamental proposition all tanners can readily assent. When the advantages of motion or force, in the sense of agitation of the fiber, are taken from the various patents which attempt to quicken the process of tanning, it will be found that there is very little left that has merit.

The writer once tried the following experiment: He took four veal calfskins, which weighed eight pounds when green, and prepared them for the liquor in the usual way. He then tied the skins together and suspended them from the end of a spring pole, so adjusted that they were covered all the time by the liquor in the vat. For the first few hours the skins were agitated by means of this spring pole in a weak coloring liquor, and were thus progressed from vat to vat until, at the end of the first day, they had reached a liquor of 16 degrees strength. At the end of four days they were fairly and even fully tanned. This spring pole was so adjusted that a slight touch of the hand would set it in motion, and as each one who passed had instructions to "lend a hand," the pole was kept in motion almost constantly, with the result indicated. This experiment should be tried by every tanner. It will attract his attention to a most important element in quick tanning. If the experment should prove as interesting as it did with the writer, it will lead to other results and conclusions. It will indicate that a most delicate touch will agitate sensibly the fiber throughout the whole skin or hide; that gentle motion is most efficacious, and that violent motion is positively injurious, as the latter purges the cells of their gelatine and prevents the plumping and final gain in weight.

The relative influence of gentle and violent motion on leather when in the handlers has been demonstrated by the action of the "rocker." Here, with the most delicate and uniform motion, it is found expedient to discontinue the movement altogether for a considerable portion of the time, otherwise the sides do not plump and take on a uniform grain, whereas by only sufficient working the most desirable result in these particulars is obtained. Many mistakes have been made, and much damage done through a misunderstanding of the action of liquor on the fiber, according to this principle. Fifteen or twenty lifts or turns per minute are ample, and these should be continued only about half the time. Besides giving rest to the fiber, this slow motion does not, to the same extent that a quick, violent motion does, turn up the liquor, bringing it into contact with the air, thus causing gallic acid to form.

These experiments, if carefully made, will also convince the tanner that the gelatine of the hide has such an affinity for the tannin that they combine much more readily than is supposed; this is made apparent by the rapidity with which the tannin is taken up while in the rocker vats. Once let a tanner be satisfied that his green stock is hungry—constantly demanding nourishment—but without the voice to make known its wants, and his financial sympathies will hardly ever allow him to sleep without the apprehension that he is neglecting his most vital interests.

Knoederer claims, and with some show of reason, that his vacuo process prevents the formation of gallic acid, and to this extent all methods that tan under liquor, in such manner as to avoid exposure to the air, either of the liquor itself or of the pelts that are in the process, should receive favor. Further than this, the method of "throwing up" the packs, it is claimed, presses out the spent liquor and prepares the vacant cells to receive an infusion of newly charged tan liquor. Aside from the disproportionate labor imposed by the old over the newer methods, the writer believes that the action of the atmosphere, both on the color of the leather and to cause oxidation of the liquor, must prove a sufficient demonstration of its impolicy.

Leather tanned while in a composed state (at rest) will

have a firmer texture than if motion is used to aid the tanning. This would probably be the testimony of the butt and bend tanners of Great Britain, and there is much in our own experience to confirm such a view of the case. But the question under review is one rather of time than of firmness of texture. None of the advocates of improved forcing methods claim, so far as the writer is aware, that they make firmer leather. They usually assume to make better time, and exceptionally claim better gains. Their improved gains they estimate to come from the saving of the waste of the gelatine, by reason of its earlier entering into combination with the tannin by their quick as against the older and slower methods. The best experience in America would direct that, after the hide is "struck through" by the most rapid combination possible, it be "laid away" for a period of weeks and even months —all the time in which its firmness of texture will improve. But if it is only required to taw (that is, strike through) the fiber with tannin and coloring matter, then the handling or fiber agitating methods are alike; for harness, upper and even calfskins this process gives very fine offal, and makes most serviceable wearing leather.

Hardly a month passes that some restive spirit does not discover, for the hundredth time, that salt and some of the sulphates, notably the sulphate of alumina, known in commerce as alum, and potassa, will in various combinations tan, or, more properly speaking, taw or preserve, hides and skins, which, after being "mooned," will make a very tough and serviceable upper leather. In this manner glove and calf-kid leather is made in a most artistic manner, and on the Continent of Europe the latter description of leather enters very largely into the consumption of the people for wear in shoes of both men and women. For dry and warm climates hardly anything more desirable could be obtained, and as it is made at much less cost than bark tanned leather, there is an excellent reason why the calf, goat and sheep skins of those countries should be so manufactured; but when these tawing processes are applied to making sole leather in this country, which abounds in bark, the tannic acid of which combines with gelatine, and is when so combined not soluble in water, then all effort in that direction is just so much wasted force.

CHAPTER XXI.

THE SPECIES AND GROWTH OF HIDES.

"HEALTHY" AND "WELL GROWN" HIDES—DIFFERENCES IN HIDES AT VARIOUS SEASONS OF THE YEAR—EFFECT OF CLIMATE AND FOOD ON TEXTURE AND GROWTH—IMPROVED BREEDS OF CATTLE MAKE HIDES THIN AND SPREADY—COLD CLIMATE MAKES A COARSE FIBER AND WARM CLIMATE A FINE TEXTURE—EAST INDIAN, AFRICAN AND SOUTH AMERICAN HIDES—THE HIDES FROM THE EASTERN AND MIDDLE STATES AS COMPARED WITH THOSE FROM THE WESTERN PRAIRIES—CARE TAKEN OF CATTLE IN EUROPE.

In a previous chapter reference was made to the structure of hides, but for quite a different purpose from that which induces the treatment of this subject here. The bison, the sheep, the deer and the goat belong to species so different from the ordinary neat cattle that no one would think of comparing their merits or defects, although the designation of "growth" will apply to all alike. A sheepskin can be a "healthy pelt" and "well grown" no less than the hide of the ox or cow. In this chapter, then, will be considered the species of cattle and their growth, in the sense in which we speak of health in the animal and vegetable world; not that all animals or vegetables are alike, but only that they have their distinctive natures, and are well or ill within those limits, according to the seasons and circumstances through which they pass.

The English tanner must be credited with the first application of the term "growth" as applied to hides. He would formerly have said a hide is "well grown" when it was healthy, plump and fine—whether it was a bison hide from the Western plains of North America, or a Spanish ox hide from the pampas of South America; but from this original and more correct designation the English tanner has come of late to speak only of plump, thick hides as well grown, and, as the term is thus used, only such hides as are suitable for butts and bends are called "well grown hides" in England. But, to use the term in a less restricted sense, the hide is well grown in the American tanner's view when it takes on its highest and most perfect nature, and then the designation applies to all hides and skins alike.

We have spring, summer, fall and winter hides, and these adjectives have to the tanner a most distinct and qualifying meaning. The hides of all our neat cattle in the spring of the year are thin, and frequently have grub holes in the shoulder and on the line of the backbone. In early summer the cattle begin to recover, and by late in June they have shed their hair, though while the hide is healthy the full growth does not come to it until September and October. The hides taken off at the latter season are thicker and healthier than those furnished at any other period. At the approach of cold weather the hair becomes long, and the hide, by sudden contractions and expansions, loses that firmness so desirable for butts and bends.

The influences, also, of *climate* and *food*, quite independent of the health of the animal, have a controlling effect in determining the value of the hide. As an illustration of the effect of climate on the fiber, the hides from Canada and from Russia are far coarser in texture, and for this reason are less valuable, than hides from the United States. Their winters are longer and much more severe than with us. This is true even when the cattle are the same in species. It is for this reason, and from this cause in part, that the Spanish hides of South America are better grown, both in the English and American sense, than are the hides of our States. The extremes of weather we have are never experienced there. The cattle feed on evergreen pastures, and are never housed, as with us. The cattle of Texas are of precisely the same species, and even of the same breed, as those of Buenos Ayres, and yet, intermediate between these, we have the cattle of the Rio Grande, of the same species but of a different breed, whose texture is far coarser, showing that climate and food cannot wholly control the structure.

The improved breeds of cattle which have been introduced both in Europe and here have done just so much to make the hides thin and spready. The improvement induces a thin skin. A "blooded horse" has a thinner skin than the old farm or plow animal. So noticeable is this that the English tanner has long since given up the idea of getting butt hides from domestic cattle. The cattle of Spain and Portugal, being of the unimproved original stock, give the English tanners the only slaughter butt hides they have. In this country we are fast improving all the plumpness out of our hides, and it will be only a few years until we, too, shall depend upon Texas and South America for all our thick pelts.

If a cold climate induces a coarse fiber, then it should be true that a warm climate makes a fine texture, and this we find to be in accordance with the fact. The hides of Africa and Central America have a fine texture beyond any other of which we have knowledge. The texture is not seemingly so much affected by food as by climate. The great droughts of these tropical regions do not to any perceptible extent control the texture, but they do affect the growth. This is illustrated by the conceded fact that the cattle of Texas that are known to die of starvation retain a fine textured hide. If it be said that the animal died before it had time to change the fiber or texture of its hide, this is not true of the growth, for the hide contracts and becomes diseased with the decay of the body. The same is true of a murrain calf or kip skin. The fiber is really closer and finer than that of a vealskin.

It is, then, probably true that where the species and breeds of cattle are the same, food affects the growth and climate the texture, and the experience of those who should be the best judges confirms this conclusion.

The "East" and "West Coast" African hides are so radically different that it may be assumed the East coast hides are from cattle brought from the Island of Madagascar, and still more remotely from Hindostan and the Chinese Empire. The hides from the West coast, bordering on the Atlantic, are substantially the same as those from our cattle, only dwarfed by imperfect food and culture. Australia, on the other hand, shows more direct signs of contact with Great Britain, whose colony she is. The cattle and hides from this province, although produced under the same climate as that of many other islands of the Indian Ocean, are distinct in character from all the rest, and when made into leather may pass for the hides of well grown neat cattle, equal to those of Great Britain or the United States. But the hides from the smaller groups of islands lying in the same ocean, as well as from China and Japan, are for the most part from cattle that are popularly known as East India cattle, so small in frame and delicate in outline, as compared with our cattle, as to have received the designation of "kips" rather than hides from full grown cattle, as they are. This designation covers that large class of East India kips from which British tanners make a most serviceable light upper leather, and the imports of these kips at London and Liverpool amount to about five millions yearly. The cattle from which these hides are taken are quite uniform, and when full grown about the size of our yearlings, but are much more delicate in limb and feature.

The part which this East India hide product is to play in the future supply of the world is as problematical as the question which is now agitating our country about the labor supply from that quarter. When we bring ourselves into new and friendly business relations with 400,000,000 of people, occupying a vast territory, and with a diversified climate and soil, we may well hesitate about our conclusions—and this Centennial year joins us to China and Japan by ties more close and direct than it has been the fortune of any of the nations of Northern and Western Europe to enjoy through all their previous intercourse.

The influence of climate also finds a notable illustration in the character of hides and skins taken off in the new Prairie States of our own country. All concede that the species, and even the breeds, are the same as with us in New England and the Middle States, and yet the texture of their hides and skins is much coarser; particularly is this true of the calf and kips of those States. The reasons for this conceded fact are as follows: The cattle are seldom housed, even in extreme winter. The farmers have no barns, or even protecting sheds, and the cold so overcomes the cattle as to seriously affect their growth. This may be noticed in their calves. A calf dropped by a cow in the Prairie States will hardly stand on its feet until eight or ten days old, while at the East they will caper and play within twenty-four hours. The calf from a well-cared for cow, at the East, will mature and its meat may be eaten in three or four weeks, while at the West from six to eight weeks is required. This same want of strength and growth is as observable in the pelt as in the meat. The difference between the "drop" or "deacon" skins of Ohio and those of Northern New York should, according to the theory heretofore given, be in favor of those of Ohio, for there is fully five degrees of latitude against New York, but with us the cattle are carefully housed during the fall and winter months, and this care more than compensates for the difference in climate, and results, as all tanners fully understand, in giving character to the calf pelts of the respective regions.

In accordance with this perhaps too hasty conclusion in regard to the effect of care and culture of the neat cattle upon their offspring, the question arises whether the conceded fineness of texture and perfect growth of the German, French and Swiss calf may not in some measure be attributed to the almost humane care which is bestowed in these countries upon their domestic animals? And if this fineness of texture extends to the skins of the offspring, may not the hide of the parent animal be measurably affected? These may be mere speculations, but they are based upon considerable observation, and have received the sanction of cattle breeders of much renown. The cows of Holland occupy the warmest and best portions of the family dwelling, and are cared for with as much tenderness as any of the inmates. It may be said of the domestic animals of all Central and Southern Europe that they are cherished and cared for in a much larger measure than the human family, supposed to be their lords and masters. May not this culture and care have something to do with that superior growth and texture which seems to characterize their calf, goat and sheep skins? This discussion may be too general to satisfy tanners that breeds, climate and food control the value and qualities of hides and skins, and yet they must admit that the theory here broached is in accord with common observation.

CHAPTER XXII.

FRENCH AND GERMAN CALF AND KIP.

WHERE OUR IMPORTED STOCK COMES FROM—CAREFUL ASSORTING OF THE RAW STOCK TO INSURE UNIFORMITY IN WEIGHT, SUBSTANCE, AND GENERAL CONDITION—SOAKING AND MILLING—BREAKING THE NERVE—LIMING—BATING AND WORKING OUT LIME—COLORING AND HANDLING—LAYING AWAY AFTER WORKING—STUFFING—DRYING—SLICKER WHITENING—BLACKING AFTER THE STOCK IS CUT OUT USUAL IN EUROPE—VEGETABLE OILS USED INSTEAD OF FISH OILS—DEFECTS IN FOREIGN CALFSKINS—STEADY IMPROVEMENT IN AMERICAN CALFSKINS.

In what is here said of French and German methods of tanning and finishing calf and kip skins, the statements may be regarded as applying also to the production of this class of stock in Switzerland and Austria. Indeed, contrary to the general impression on this subject, these countries are about equally represented in the manufacture of this description of light imported stock, commonly styled by consumers "French" goods. We seldom hear of "Swiss" calfskins, although Mr. Mercier, of Lausanne, Switzerland, is the largest manufacturer sending leather to this country. The common designation of "French," as applied to all imported skins, arises from the fact, probably, that American agents for the purchase of these skins reside mostly in Paris, and ship largely through French ports. This consideration is important in this connection only as showing that a characteristic and almost uniform system or method of manufacture prevails in all the countries mentioned, and the goods they produce may, therefore, be treated as one manufacture.

There is one other preliminary consideration which should be here stated, viz., only the best goods of manufacturers on the European Continent are sold to this country or to Great Britain—that is, only those parties that manufacture calf and kip as a specialty, and in large quantities, ever think of selling their products away from home. Of the ten thousand and more tanners of these countries probably not more than fifty find a market in the United States for any considerable portion of their stock. The local or country producers of calf and kip in all these countries labor under the same disabilities as do the large number of small tanners in our own country. They are not able, from their limited facilities, either to obtain skilled labor or to make the same selections as do those tanners who make a specialty of this class of goods, and do the business in a large way, and hence it is that the *people* of Continental Europe generally wear no more artistic calf and kip than may be found in the shoes made up for the ordinary wear of a large portion of our own population.

In respect to weight, substance and general condition, the skillful foreign tanners take especial pains to see that the skins of each pack shall be as nearly alike as possible. In this respect their practice does not differ from the theory of our largest and best calfskin tanners, but by reason of the large unmanufactured stocks which they usually carry they are enabled to put their theories into practice more thoroughly than it is possible for our tanners to do. Their calfskins are for the most part dried without salt and folded on the back; the selections for weight are then made, and the skins are packed in bales. This is done by the factor or dealer. When the goods reach the tanner he still further assorts and then packs them away in cool lofts where they await his wants. How different these circumstances from those presented to our large calfskin manufacturers, who buy all their stock in a green salted state from the 1st of May to the middle of July! The condition of the green salted skins as handled by our tanners taxes their best energies; to keep them from salt pricking and decaying is very difficult, and the tanner is forced to hurry them into the tanning process with indiscriminate haste.

When skins in all respects alike go into the soak, they are equally affected by the water, and are ready for the mill at the same time, thus enabling packs that are commenced together to go through to the end without separation or division. This uniformity also greatly aids the general economy of the manufacturing by keeping the tanner informed of the amount of product which he is realizing from his raw material, for each pack or series of packs will indicate to him his loss or profit. Aside, however, from the general merit of this classification at the start, it may be asserted confidently that no tanner can make uniform and good stock without such selection, even with the greatest care in the after handling.

The treatment necessary at this stage of the process will be greatly influenced by the condition of the stock. If the skins are green salted, as with us, a very slight milling will suffice; but if they are flint dry, as with some of our small tanners, then they should be thoroughly softened. But any forced softening must be preceded by a certain amount of soaking, which in turn will greatly depend on the condition, whether green, green salted, dry salted or absolutely flint dry. The soaking process will, therefore, extend from a few hours to a few days, depending upon the condition of the skin and the state of the weather. If the skins are flint dry, they should not be milled until after ample soaking, and even handling in the soak; but the first milling should be gentle, otherwise the grain of the skin may be cracked. This milling should be only for a few moments, and with a full mill, so that the large body of skins will present a soft, yielding mass to the action of the hammer, if the softening be accomplished with the ordinary fulling stocks. If a revolving wheel is used, then the same care is necessary that the skins may not have too violent a pounding on the inside wooden projections. After the first slight milling the pack may again be returned to the soak or thrown up in piles. The action of the water in the second soaking will be much more rapid, and great care should be taken that the grain does not "prick." After the pelt has been in this manner made as soft as water and ordinary milling will accomplish, then we notice the first peculiarity in the methods of the French and German tanners, as compared with the practice here, which consists in the thorough breaking of the nerve.

The nerve to be broken is a description of interlacing fiber which holds the animal tissue. This nerve is located transversely on the flesh of all pelts. It may almost be said to make a part of the flesh itself, and lies immediately under that loose, fleshy tissue which curriers shave off before they reach the pelt proper, on which they form their waxed surface. The office performed by this nerve or tissue is, by its contraction and expansion, to hold the pelt close around the animal. It forces the pelt to conform its shape to that of the animal whose body it covers. When the animal is fat and well rounded out it expands, and when poor, sick, or thin, it contracts. It is this nerve which induces "bagging" in all leather, notably in sole leather, while in harness and upper leather, which is stretched by shaving and scouring, much of the contracting force of this nerve is destroyed. But something more and beyond the avoidance of "bagging" is required in the preparation of calfskins. The nerve must be so completely severed or broken that the whole pelt will feel not only soft, but *actually pulpy*. This effect can only be produced at present by hand labor. No machine has yet been devised to do it, although it may be predicted that, when the great utility of the work becomes appreciated, a labor-saving machine will come forth from the brain of some American to perform this labor. The hide-working machine of Mr. Henry Lampert, of Rochester, N. Y., when improved, may answer the purpose, or suggest the way to some other and better machine.

If the skins are dry or dry salted there is no difficulty in determining when this nerve is sufficiently severed; but if they are green or green salted, unless more than the usual care and conscientiousness is exercised on the part of the workman, the nerve will remain unbroken, and will hold the whole fiber of the pelt firmly during the entire after tanning process, so that it becomes difficult if not impossible ever afterward to make a yielding and elastic substance from the skin thus treated.

Assuming that we have succeeded in breaking this nerve, and have brought the pelt to a soft, pulpy condition, the next process will be liming and unhairing. The skin has been brought to a condition in which it will readily take the lime, and four or five days in a moderate lime water will neutralize the animal grease, and soften and swell the roots of the hair so that the latter will come off with the most inconsiderable labor. No more lime need be used than sufficient to accomplish these two objects, although the practice in this respect is not uniform; the statement is made rather on a knowledge of the American experience than from personal observation as to the practice of foreign tanners.

When the hair is removed the skins should be thrown into

the mill or wheel and washed for a few moments. If in warm weather, water at a temperature of 60 degrees should be used, but if the weather be cold, the removal of the lime will be greatly facilitated by using water a temperature of from 80 to 100 degrees. This warm water and milling process must be conducted with judgment and care, but when so carried on it is perfectly safe, and considerably shortens the process of depletion, besides saving the expense of other depleting agents, usually called "bates."*

The lime must be worked out before the skin goes to the handler. In the practice of some of our American tanners, the acid of the liquor is relied upon to overcome a portion of the lime which may not be readily removed, very much as is done by some of our sole leather tanners; but, according to the best French process, the lime must be thoroughly worked out before the skin goes into the handlers, and all the working is done on the grain rather than on the flesh side. The result of this treatment is that the skin, when properly prepared for the handlers, is so depleted that the pelt of an ordinary eight-pound green veal skin can be drawn through a ring two inches in diameter.

After all that has been said or experienced by tanners, it is probably true that much depends on the kind of liquors which are applied in the handlers. Skins depleted by bates in the ordinary way, and subjected to sweet liquors in the handlers, will, undoubtedly, give a finer flesh and a tougher fiber than when otherwise treated, but whether the skin will be as plump and as full in the offal may well be doubted. If, however, there is any lime left in the skin, sour or acid

* Some of the best and largest tanners of Switzerland depend very largely on the sour acid liquors of the "larch" bark, both to form their color and at the same time deplete the skins. One of the most successful calfskin tanners of Switzerland told the writer that he did not consider it safe to deplete with ordinary bates, and then trust to sweet liquors to color and tan the fiber.

liquors must be so far used in the coloring and early tanning process as to overcome the last particle of lime, otherwise the tanner will have a coarse flesh and spongy fiber that will not "carry" stuffing in the finishing process.

The methods of coloring and handling followed by the tanners on the Continent of Europe, so far as they came under the observation of the writer, would probably fail to give satisfaction to American tanners. The handling there is done in vats, with old sour liquors, replenished with either "larch" or partially spent "oak" tan bark. About one-half of a pack is put into a round or square vat, and a man is constantly employed in raising the skins to the surface by means of a pole, very much as was the practice of our tanners fifty years ago. Other tanners, more advanced and doing business in a larger way, have a revolving wheel in the top of their vat, constructed and operated very much after the style of our so-called "England" wheel.

The process of merely coloring and handling soon terminates, and then the skins are paired and laid away, grain to grain. The object of this grain to grain laying away is two fold—first, to tan the skins from the flesh side; and, second, to keep the grain smooth and tough. Both of these objects are accomplished by this process, and, next to the breaking of the nerve in the beam house, this may be considered the most important innovation upon our method. The skins are paired according to size, and carefully laid pate to pate and butt and butt, a matter which is not difficult in a pack that has already been selected with great care to secure uniformity in weight and substance. A man will size and place these skins grain to grain almost as fast as he could handle them in any other way. When so placed and held by the hind shanks with both hands, they are passed to a man in a deep round vat, from which the liquor has been drawn. This man first takes a shovel full of spent or partially spent bark from one attendant, and then a pair of skins from another, laying first bark and then skins, and thus alternating until the tank or leach is nearly filled, walking around on and packing down both skins and bark, taking care to fill up all intervening spaces with bark. These vats are usually round, being six to eight feet deep, and eight to ten feet in diameter, about ten times as much space being occupied with bark as with skins. When the vat is filled to within a foot of the top, water or old liquor is run on until a covering is obtained, and this first layaway is continued for from ten to thirty days. The second, third, fourth, and even fifth, layaways follow at intervals of from twenty to forty days, each time, however, with more of an admixture of fresh strong bark with the old and partially spent tan, and with an increased strength in the liquor run on, but so far as the writer has observed no new bark is leached and the decoction resulting put on the packs.

The result of this practice is to tan or color through the stock in four, six or eight months, and this is done from the flesh side mainly, the grain of each skin being at all times covered and protected by the grain of its mate, without any bark or bark liquor intervening. The effect of this close proximity of animal fiber is to prevent anything like the formation of old grain, and where such grain does form it is so soft and yielding as to be readily "pulled out" or "boarded out" without scouring in the process of currying. The grain, when the skins are first separated, indicates just such a condition or appearance as would the grain on a pack that had been allowed to sweat in piles, but as the grain surfaces have been absolutely excluded from the air during the whole process they are without stain, of the most delicate color, and extremely fine and pliable.

The tanning processes being completed, the bark is thoroughly shaken out as the skins are taken up and laid in piles to drain and "sammy." At this period comes in the third peculiarity of European calfskin tanning, namely: The skins are worked over a small half round beam, on the flesh side, with an ordinary tanners' worker. The object of this working is to complete and make permanent the work of breaking the nerve commenced in the beam house, and to take off any flesh which may remain. Its effect is to make the skin soft and the fiber elastic, and, being quite inexpensive, should never be omitted. It can be done by any ordinary yard hand who is faithful and honest, but there must be no slighting.

One of the characteristics of foreign workmen is that they are faithful to instructions; any manipulation which requires patient labor may be intrusted to them without fear of omission or neglect. So much cannot be said for American workmen, particularly those who work by the piece; they are constantly studying how to save labor, and the tanning and finishing of calfskins in a style as perfect as are the goods of our European neighbors consists of so many little things—all demanding careful, painstaking labor—that, without the greatest oversight on the part of the foreman, it is not to be expected that any large proportion of really good skins, such as the French and Germans make, will be manufactured in this country for many years to come.

After the skins have been tanned and "mooned," as already indicated, they may be flesh shaved, as before described, or, as some prefer, they may be thrown into the wheel and "rough stuffed" before being shaved. The latter method is regarded as wasteful, since the shavings absorb oil, and to this extent cause a loss; but many finishers claim that a better flesh is obtained when the skins are "rough stuffed" before the shaving is done. The writer prefers the former course.

In beginning the finishing process—the tanning and beam working on the flesh having been faithfully done—there will be very little or no loose flesh to shave off, and, while a "smooth face" must be obtained, it is desirable that no more than merely the flesh should be removed with the knife at this period. From one to three pounds per dozen may easily be lost by want of care at this stage of the process; therefore this shaving should never be done by the piece, but always by the day, and with the most conscientious workmen, for labor that is worth fifty cents per dozen can be made to cost the tanner two or three dollars by the loss in weight which the workman may cause. If, from any omission, an unusual amount of old grain has been allowed to come into the grain of the skins, it may be advisable to board them to draw out such surplus grain, or rather to scatter it, while in this rough state, but this necessity will seldom arise if proper attention is paid in the tanning process.

Assuming that all leather finishers have both scouring and stuffing wheels, it is recommended that, before scouring by hand, the skins be thrown into the scouring wheel, and all the bloom and dirt washed out. When this service is thoroughly performed, the hand labor necessary is little more than that of striking out the skin on both flesh and grain sides.

As one of the characteristics of French finishing is to leave all the "stretch" in the skins rather than to take it out, we begin our new method by omitting almost entirely the customary scouring and distending process. If the skins have been well "struck" on both grain and flesh, very much of the old liquor and water has been pressed out of them, but yet they are too wet—the fiber is too full of water—to allow the oil and tallow to properly enter, hence they should be partially sammied. This requires care, otherwise the oil will enter unevenly and darken the grain.

Some curriers have hand presses, and hydraulic power presses are used, under which the skins are placed to force out more water than it was possible to press out with the slicker, but the observation of the writer is to the effect that exposure to the atmosphere is the only proper way of preparing the skins for the stuffing wheel, and after they have been thus exposed to the air a sufficient time to fairly stiffen they should be taken down and placed in piles, as in this situation the moisture will distribute itself, drawing from the center to the circumference.

Unless great care is observed stuffing leather by the use of the stuffing wheel may do incalculable damage. French in its origin—its present efficiency is wholly due to American adaptation. The oil and tallow should be held in separate vessels, warmed with a steam coil at the bottom of each, and thoroughly mixed while in this warm state. When the skins are properly sammied they should be thrown into the stuffing wheel, the proper amount of stuffing added, and the wheel set in motion. Ten to twenty minutes will suffice to mill this mixture of oil and tallow into the skins, but the wheel should be allowed to run long after the stuffing is outwardly absorbed, since this fulling process will work the stuffing into the center and thicker portions of the skins. It is possible to put in an excess of oil and tallow, and this is quite common, but such excess may be got rid of in one of two ways, as follows: First, carefully pile the skins on each other as they come out of the wheel, and by their own weight the excess of oil will be pressed out; or, second, throw each skin on a table and gently work out the excess by means of a steel slicker, using judgment to press equally over the wholly grain surface. If done with sufficient care, this action will relieve the flanks, pates and bellies of any undue proportion of grease, and the skins may then be hung up to dry, either from tenter hooks or over sticks.

Although skins freed from the acid of the bark and saturated with oil are not so liable to be affected in their color by means of light and air as ordinary tannages, yet they are sufficiently so to require care in these respects. The drying loft should, therefore, be so constructed as to enable the attendant at all times to regulate the amount of light and air which should enter, and the drying should be slow. It will always be safe to allow night air, which is without light, to enter. In the winter months great care should be taken to avoid excessive stove or even steam heat, the tendency of all artificial heat being to make the leather harsh and bring the grease to the surface.

After the skins have been stuffed and dried in the manner indicated they may be packed away in piles of 200 or 300, and allowed to remain for a great length of time, as they will improve all the while. It is in this condition that the large stocks of Europe are carried, goods made out of season or which are unsalable from any cause being thus held over for years.

The process of slicker whitening is preferred to the old method for three excellent reasons: 1. Less of the weight is taken off. 2. A better (smoother) face is left. 3. A workman can do more work with one of these tools than with the usual whitening knife.

While the skin is in the dampened condition, which has been brought about both for the convenience and more effective work of the slicker whitener, it is thrown upon a table covered with leather, grain up, and by means of a long "grain boarder," held firmly by the whole arm, the entire skin passes under the operation of the board, but especial attention should be paid to the neck and pate, or wherever else an old, elongated grain may be seen. The effect of the boarding will not be to make the grain entirely smooth, but so to scatter it as to render it less observable. It is no part of the duty of the French currier to make a smooth grain; indeed, this is just what he wishes to avoid; but it is part of his art to make a smooth and fine flesh surface with a grain so loose and yielding as to adjust itself readily to the operation of the crimper.

Contrary to our manner of proceeding, the calf and kip skins in Europe intended for boots are not blacked on the flesh until after they are crimped on the boot trees. They are bought, sold and held in the "russet" state. While in this condition, (cut in forms), they are carefully inspected, and wherever small cuts, hacks or flaws of any kind are observed on the flesh they are buffed or shaved out. The effect of this operation is to make the skin slightly thinner at the defective point, but in cutting in forms these defective spots are so located as not to injure the wear, and at the same time to escape the notice of the consumer. This work is done by boys or women, who, with broken glass or thin turned steel edges, pare down the lips of the cuts or hacks so carefully as to hide them from view. After this is done, the fronts are blacked, if a special trade demands it, but usually the goods pass to the manufacturer in the russet state, to be by him blacked after they are made up.

Before crimping the boot front, the grain at the instep is buffed from each front. The reasons for this are as follows: 1st. If the grain is left on the crimp would pucker it, and render it rough to that portion of the foot; 2d. By removing the grain the crimp is effected with less labor, as that portion of the skin is made more elastic; 3d. The tendency of the process is to force more volume of fiber just at the turn than elsewhere; but 4th, and mainly, it softens the skin at the point where there is the greatest friction and pressure on the muscles of the foot. It is quite notorious that buffed leather is much softer than that in which the grain is left on, and it often happens that men with tender feet require their calf-skins to be buffed over the whole grain surface before being made up. The reasons for the practice of buffing so much of the crimped form as covers the instep will therefore appear sensible.

The leather finishers everywhere south of Great Britain use very largely and quite uniformly the vegetable oils, mixed with tallow, instead of the fish oils used by us. This difference alone is quite enough to account for much of that softness and elasticity of fiber so much admired in their leather product. The degras of France is about the equivalent of the sod oil of Great Britain and the buckskin oil of America, but there is this difference between them: Vegetable oils make the base of the degras while fish oils make the base of buckskin and sod oils. Palm oil is very largely used by the calfskin manufacturers of France and Switzerland; cocoanut oil, castor oil, cotton seed oil, and, by parity of reasoning, linseed oil, may all be serviceable, but fish oils should never be used. They cause leather to gum, and their whole nature seems destructive to the fiber.

There are three general defects which should not be overlooked in the French methods of manufacture:

First—They do not properly trim their skins; the offal usually found on French calfskins is so thick and coarse as to be perfectly worthless. That a people otherwise so economical in their methods should have fallen into this defective manner of trimming, if, indeed, it can be called trimming at all, and continue the practice for so many years, is a matter of surprise to all thoughtful persons who have to do with the manufacture.

Second—By the French method of tanning and finishing the leather made is less impervious to water than that made by either the English or American methods, rendering the former only suited to dry climates, or for wear in large cities, where it is not exposed to wet or damp soils.

Third—The starved nature of their tannage—the very effort they make to render their leather soft and yielding in texture—deprives the shoulders, bellies and flanks of that gelatine which, when combined with tannin, enables the manufacturer to cut his stock to the very outer edge, thus making every square inch serviceable.

It is in great measure from the want of fineness in the offal of foreign goods that our shoe manufacturers are substituting American for French calf. So long as our people wore boots the coarse offal of the French stock could be run up into the legs, but now that we have become a nation of shoe rather than boot wearers, the change has induced a reversal in the public judgment in regard to the economy of the French method of tanning and finishing calf.

Perhaps also it is due to the American calfskin manufacturer to concede that he has greatly improved in making a more elastic fiber, while he has retained the fineness of the offal and closeness of trim. It must be within the observation of all leather consumers in this country that both calf and wax upper leather have undergone a great change for the better within the past few years. A few more strides forward and the upper leather of America will stand on a par with that of the best productions of the most advanced nations of Europe.

In respect to the proper classification of skins the French and German tanners have always had the advantage, and this is especially the case with those large manufacturers sending their stock to this country, who produce as many as from 2,000 to 3,000 dozen per year, and in exceptional cases vastly more. This enables them to select male from female skins, and make several qualities and selections inside these general classes. Until the calfskin tanners of America shall imitate the French tanners in this particular they will labor under present disabilities, but the energy and forethought which has brought us so near competing on equal terms, will, there can be little doubt, carry us through to that desired period in our international intercourse when our tanners shall no longer fear the effects of the removal of all duties, and a free exchange in all the leather products, to the mutual advantage of all the nationalities of the world.

CHAPTER XXIII.

GRAIN AND BUFF LEATHER.

SPLITTING MACHINES—MAKING SPLIT LEATHERS FROM GREEN HIDES OR FROM TANNED LEATHER—EVERY KIND OF NATURAL GRAIN SUCCESSFULLY IMITATED—STRENGTH AND DURABILITY OF SPLIT LEATHERS—THEIR INTRODUCTION TO EUROPEAN CONSUMERS—ESSENTIALS TO BE CONSIDERED IN THE MANUFACTURE OF GRAIN AND BUFF LEATHER.

Commencing with the successful introduction of the union splitting machine, about the year 1830, leather made from the hides of neat cattle has been split down to a comparatively thin substance, and the "grain splits" have been used extensively for shoes, as well as trunks, harness, etc. With the improvements made in that machine in later years, very safe and good work has been done in splitting the ordinary cow hides into both stout and light grain leather, while the best of the flesh splits has been used extensively for boot backs and shoe quarters, and the middle splits for trunks. More recently, or about 1860, the "endless belt knife splitting machine" came into use, and it splits with even greater precision than the union machine. From its high cost (about $1,000), as well as from its complicated structure, making it liable to get out of order, comparatively few of these machines are in use. When it is new and in good order, and handled by a competent workman, an ordinary cow hide can, with it, be split into three, and even four distinct parts, with as great precision as the sheepskins of

England are split grain from flesh. The merit claimed for this endless belt machine is that the fiber of the hide is cut without strain, while it is alleged that with the union machine considerable force is required to draw the side through. This objection had much force with such machines as were constructed early in the history of splitting leather, but with the improvements since introduced, there is no perceptible disturbance of the fiber, and no greater power is required than can be applied with the attendant's hands. These union machines are now so perfect and durable in their construction that they may be said to be indestructible by ordinary wear, and, with average skill and care, grain leather of the thinnest description can be split from the ordinary neat hide. An illustration of the union leather splitter will be found in subsequent pages.

So much has been said in regard to the machines and facilities for spliting that we may consider the intrinsic merits of the grain and buff leather which has for the past few years come into such universal use in this country, and is now, by its cheap and excellent qualities, commending itself to all the countries of Europe. A clear distinction must be kept up between the process of manufacturing these grain and buff leathers from the green hide and those made from tanned leather. Our curriers are too often tempted to split down fully tanned leather into grain. Such efforts bring into disfavor all grain leather, since, from the very nature of the case, the grain must be tender when fully tanned first and split afterward. It is a great misfortune that grain leather made in this improper manner cannot be discriminated against by buyers, as it is almost sure to give dissatisfaction when made up into shoes; but, tempted by a concession of one or two cents per foot, manufacturers are too apt to "try just this one lot," and that one lot, however small, carries discredit to a whole class of goods. These objections apply only to light women's grain. Men's boot grain, which is split to weigh seven to eight ounces to the foot, will have substance enough to hold against any strain.

But the object of this chapter is to consider the intrinsic qualities of modern grain leather, manufactured in imitation of calf, goat, seal and even hogskins. By far the largest portion is manufactured as "pebble" or "goat." The grain is made coarse or fine, to suit the taste of the purchaser. Indeed, with present appliances, an order can be filled at short notice with just such devices imprinted on the grain as may be desired, and without additional cost to the purchaser. Probably there is no animal that lives whose skin cannot be imitated, so far as the external appearance is concerned, and this can be done by a machine which will duplicate the impression indefinitely. The effect of the production of this description of goods on the manufacture of goatskin morocco in Europe is just beginning to be felt. In this country its influence has been very perceptible for the past five years.

The question which consumers desire to be satisfied upon is as to whether it is *tough*, so that it will wear without tearing or cracking. In respect to "cracking" there can be no hesitation in saying that, from the nature of the pelt and its treatment, it is not more liable to crack than East India kips, goat or seal skins. It will also take as durable a color, *i. e.*, it will not turn "foxy." But is it as lasting in general structure? Can a part of a pelt be made as strong as the whole? The answer is both no and yes. In the bending of the fiber of the leather so frequently as is made necessary on a man's or woman's shoe, the stock is a great deal more liable to crack if it be of thick leather than if thin, but if the thin leather is made from a pelt that has its whole structure, both flesh and grain, and all intermediate fiber, it must be more serviceable than when this structure is broken, as is the case with these split grain leathers. But suppose we admit that this grain upper is less tough than goat, seal or horse leather, if it is *sufficiently tough to wear out two pairs of soles*, as is affirmed by our experience, and can be offered at thirty per cent. less price than all competing leather, then will it not be pronounced a success, and will it not largely take the place of these leathers throughout Europe? That it will greatly affect the consumption and price of goat, calf, seal and horse leather throughout the whole world there can be no doubt; whether it shall ultimately come into general use for women's and children's shoes will depend upon the honesty and fidelity with which it is manufactured. It is now almost exclusively made in this country, and because we have the machines to split successfully the hides. When the leather manufacturers of Germany, Austria, Switzerland, and even France, shall turn their attention to the production of this leather they will successfully compete. Outside of our machines and our trained hands for splitting we have no advantages over those countries. Indeed, their cow hides are finer in grain than ours, and as their neat cattle are almost exclusively cows, there can be no reason to question a full supply of the raw material. The improved breeds of cattle in Great Britain will also give the English tanner a most suitable cow hide for grain upper.

The question, then, of substituting split grain for the skins of the smaller animals is one as broad as the two Continents. In view, therefore, of the largeness of the subject, let us consider in conclusion some of the methods and economies of manufacturing this grain and buff leather for women's and children's shoes.

First.—The hides must be free from all scratches, and from horn and hook marks on the grain. Cows sent from the West on cars will not produce hides sufficiently free from the above defects to make grain leather.

Second.—The hide should not weigh over 40 to 45 pounds, cured and trimmed weight. Leather made from old and heavy cow hides the grain is coarse, and when cut down thin is tender.

Third.—They should be worked in when as *fresh* as possible; hides that have laid in the salt for a long time are apt to have a "frized" grain.

Fourth.—The lime should be worked out thoroughly; warm water in the wheel will do this to a limited extent, but bates should be used as a last resort before final working.

Fifth.—Our best grain leather manufacturers handle and even tan by suspending the sides; this practice makes the shoulders and flanks very fine. Whether it adds to the toughness of the grain may be questioned; but for buff leather it is indispensable, and even for grain it presents so many advantages that it may be doubted whether any manufacturer can afford to dispense with the practice.

Sixth.—The side is *colored* and *raised* before splitting; that is, the sides are handled on sticks (suspended) until they begin to feel firm and put on the appearance of leather. This usually occupies ten days in a weak liquor, when the sides are taken out, drained and stiffened, (by being hung up in the loft). They are then split. Usually there are two splits taken off, one main split, and one junior or small one, which covers the kidneys, but which does little more than flatten the side. If these freshly split parts are thrown back into a strong liquor, or even one of moderate strength, without wheeling or brushing, they will become "crusted" over, and will not take the tan; but if thrown into the wheel for a few moments, and run in a weak sour liquor, then all difficulty on this score is avoided. The old method of *brush-*

ing with a stiff brush is now considered too expensive, particularly as the wheel answers the purpose. After this wheeling the grain portion is hung in the vat and the split portion is laid away with bark.

Seventh.—The liquor in the vats is changed, but not the sides. Some run off, say one-quarter or one-third of the liquor into the junk, and pump back on the newer packs, replenishing the head packs with new liquor from the leaches, while others *press round their yards* on the same principle that the press leaches are run. In either case great care is taken to feed the leather both slowly and uniformly, so that the grain is kept soft and full without "drawing." The effect of this method is to give to the grain portion of the hide the same amount of tanning that would be given to a whole skin, *i. e.*, it is tanned from both sides, leaving in the center an imaginary line of partially tanned or only colored fiber. This imparts a toughness that distinguishes the leather from ordinary grain leather made from fully tanned hides. By reason of the thinness of the grain portion very weak liquors can be used, and yet the tanning is completed in thirty or forty days. It may seem anomalous, but it is nevertheless true, that tanning makes the fiber of hides comparatively tender, or less tough, and, beyond tanning sufficiently to preserve the gelatine, the further filling of the fiber for upper leather should be avoided. The weak liquor process, then, which these grain leather manufacturers are enabled to pursue, is well calculated to make a tough fiber.

Eighth.—Another advantageous result of splitting the hides intended for grain leather while green is that we get rid of the less elastic fiber of the flesh, including the nerve, and the grain will stretch until it comes to its proper bearings. This nautical expression conveys the idea perfectly. The fiber must be distended before it goes into the shoe, and it is better to let it distend itself while green than to tan it and stretch it afterward. This point may seem obscure and even doubtful, but the writer is satisfied that there is much in it. To illustrate—Why is it that damp or wet leather is much tougher—will bear greater strain—than that which is dry? Simply because all the fiber pulls evenly when wet, but when dry there is not the same uniformity. For the same reason, a side of rough leather is less tough than the same side after being scoured and set out, and when grease is added still greater strength of fiber is secured. It is found that, in the manufacture of wire rope, greater strength is secured by laying the wires side by side in a bundle, rather than by twisting them up as in a rope. Now it may be assumed that the more equally and perfectly the fiber of the hide can be distended the greater will be the aggregate strength. Whatever may be the rationale of this subject, the fact stands conceded that grain leather manufactured in the way indicated is much tougher than when made in the old way.

The scouring, stuffing, blacking and embossing, is conducted much in the same way as by the old method. Each of these processes are greatly facilitated, however, by having a uniform substance and tannage to treat. The bloom and extractive matter, for instance, can be washed out of these grain sides in the wheel so completely that they require little more than smoothing off before hanging them up to partially dry before stuffing.

In all trades a better result is obtained when a proper division of labor is observed, and this is notably true in the manufacture of this description of leather. We have now several large manufactories devoted exclusively to the production of grain and buff leather. They keep in mind the result they wish to produce from the beginning, and only buy such hides as will suit their purpose. This regard to true economy is not new in Great Britain or France, but is here, although the indications are not few that our tanners are waking up to a better conception of the true economies of their profession. With this changing condition of the trade, it may be possible that the morocco and calfskin tanners of our country have greater occasion to fear competition from our home "pebble grain" manufacturers than from foreign importations of their own goods.

CHAPTER XXIV.

CURRYING AND FINISHING.

THE STUFFING WHEEL AND HOW TO USE IT—TO PURIFY AND CLEANSE DIRTY GREASE—HOW TO MAKE STUFFING—FLESH BLACKING—FLOUR AND SIZE PASTES—HARM THAT MAY BE DONE BY DEPENDENCE UPON RECIPES—DAMPENING LEATHER BEFORE AND AFTER APPLYING OIL AND TALLOW.

In a conversation with Mr. CHARLES KORN, (than whom no man in America is better informed on the philosophy and practical methods of dressing and tanning leather), the writer asked him to commit to writing his views on the general subject of his profession. He said: "Impossible! It would only mislead; no man can know how to finish leather from any written statement, however clearly expressed." To the request to write out for publication his methods of making oil and soap blacking; also his methods of compounding size and paste, clearing oil that is impure, etc., his general answer was that it was not practicable. Nothing short of practical experience, adapting each mixture to the particular kind of leather had in hand, would ever make a successful and workmanlike job. And he was right. As well might a landscape or portrait painter attempt to give directions how to paint, as for a currier to attempt to give written instructions how to finish leather.

There was one subject that Mr. Korn did discuss with considerable warmth. Several efforts having been made to patent the stuffing wheel, which he introduced and gave to the trade, without claiming a patent, he said that all attempts to heat up the air, or the oil and tallow, by inserting coils of steam pipe inside of the wheel, or by introducing heat artificially in any way, were useless. His method of mixing his oil and tallow had been uniform and always successful. The oil and tallow should be heated separately and mixed, before putting into the wheel, in such proportions as the nature of the leather and the season of the year require; in the winter months less tallow and more oil; in the warm, summer months, more tallow and less oil—both seasons varying from an equal quantity of each, according to the circumstances. It is desirable to get as much tallow in the leather as possible, without making it hard or causing it to "spew." If the oil and tallow congeal and attach to the sides and corners of the wheel, as is often the case, a jet of steam from a pipe applied to the inside of the wheel for a few minutes will so soften it as to make it limpid, when it will unite with the fresh oil and tallow. When once the wheel is in motion, the friction occasioned by the rubbing of the sides or skins against each other will produce heat enough to keep the oil and tallow from congealing, particularly if the atmosphere is moderately warm in the shop, as it should be always.

It is a remarkable fact that, long after this wheel was introduced into public use here, a Scotchman visiting his native country, and seeing that his people had not heard of this improvement, actually sold the exclusive right to use the machine to a very large and enterprising manufacturing firm in Leeds. With the free intercourse between the leather trades of Great Britain and the United States at this time it would hardly be possible to repeat that operation.

The slicker whitener, which has now come into almost universal use, is to be superseded, as many think, for all work except calfskins, by one of the three newly patented whitening machines noticed in subsequent pages.

The purifying and cleansing of dirty grease before it is

put on the leather is accomplished by applying heat. Usually, the sun on a warm day will clarify oil that is exposed to its rays. But the more speedy and effective way is to place the dirty grease in a barrel, and, for every thirty gallons, apply about one quart of sulphuric acid; stir it very thoroughly, and afterward steam the oil until the whole fairly boils. The condensed water from the steam, aided by the acid, will disengage all or most of the impurities of the oil, and carry them to the bottom, when the whole is allowed to stand and settle. The dirt may be drawn by a faucet from the bottom, or the oil may be carefully dipped off from the top. In this way the writer has seen railroad grease, as black as ink, so purified as to make very serviceable oil. This railroad oil is obtained from the cotton waste used in the boxes of the axles, and is probably the most unsightly oil offered the currier, and yet it can be made serviceable in the way indicated, both as to cleanliness and color. If the acid should prove objectionable then pulverized chalk, stirred in the oil and allowed to settle, will carry down and off its objectionable features. The currier will seldom have oil as bad as this to handle; generally he has only the common "*head fats*" and scrapings from the tables, which will yield to the treatment above indicated most readily.

A small tank or hogshead may be kept standing in a convenient place, with one head out, and fitted with steam coil in the bottom; it should have, also, two or three faucets at short distances apart, near the bottom, from which all impurities may be drawn; this apparatus will prove a most useful one in the purification of all greases. If steam is not condensed in the oil, but the heating is done by a close coil, then water should be freely poured in with the oil, and all boiled together.

The making of stuffing with ordinary oil and tallow is well

understood, but of the use of degras or sod oil in connection with oil and tallow less is generally known. Mr. Korn furnishes the following suggestions: Melt the tallow, and let it stand until it is partially cool. If, however, degras or sod oil be mixed with the tallow, it must be stirred in while the tallow is hot; the oil added may be put in while the tallow is cooling. A little water with the sod oil and degras will lighten the color very much, but the oil must be added after, never before the degras.

Flesh blacking is made either with lampblack and soap or lampblack and oil; if "soap blacking" is used, oil is freely applied afterwards to fasten the color and body of the black. The principle advantage claimed for soap blacking is that it fills the flesh with a better body and covers defects which show through if oil and lampblack alone are used. The disadvantages are that when this leather is crimped for boots, the blacking washes off and leaves a coarse surface, while the alkali in the soap will, if allowed to lie for a long time, more or less destroy the grease and impart a harsh feeling to the leather; pure oil blacking will grow softer by age. So far as known to the writer soap blacking is now used in no other country, and is at present used here only in the modified form to which attention has been called.

Flour paste is used first after the soap blacking. It is made of flour, with soap added—say to every pail of flour use two pounds of hard brown soap, which is to be boiled with the paste. Some add tallow also—for every pail of flour paste using a piece as large as an egg.

Size paste is made as follows: Dissolve four ounces of glue in warm water; add a small piece of tallow, say half an ounce; dilute in water until the proper consistency is obtained to spread easily with a sponge.

The writer refrains from giving further suggestions on these topics lest he should subject himself to the criticism with which Mr. Korn closed his observations: "Writers of books and lecturers are mere theorists, and are seldom good workmen." A small treatise on Tanning and Currying, by "a Maryland tanner and currier," in addition to other leading suggestions, contains "twenty-five valuable recipes," on which hang most of the principles involved in the tanning and finishing of leather. Should the writer attempt to imitate his example, and lay down the absolute rules of action governing all cases, giving recipes in a dogmatic way, it would probably do more harm than good. For instance urine, until within the past twenty years, has been the only known solution to kill the grease preparatory to the blacking of the grain of leather, and all recipes were based on this chemical agent. Now, the chemist who suggested that any other alkali would do this service just as well and a great deal more uniformly saved the currier one of his greatest nuisances, and insured a much improved result. At present the use of soda ash or sal soda takes the place of the "sig." barrel. The recipes above mentioned are full of antiquated notions of this kind, which it will be well to forget as soon as possible. What we want to know is the chemical characteristics of all the agents we employ. Then we shall know how to substitute the one for the other, and learn to employ those that are the most economical as well as serviceable.

In closing this chapter the writer would call attention to a new and, to some extent, revolutionary idea in regard to the stuffing of leather. It has always been held indespensable that leather should be partially wet or dampened before the oil and tallow was allowed to enter the fiber. If applied when too wet the oil would not enter, and if too dry the fiber would be "burned;" this has always been the popular idea. No miner would think of oiling his old boots, no farmer his old harness, no mill owner his old belt, without soaking in water and preparing the leather to receive the grease. It is the opinion of the writer that no greater fallacy ever prevailed. Worn out leather—that is, leather that has lost its vitality by the evaporation of its grease, should be replenished by having grease applied to it while in a *dry state*, and after the oil or grease has entered and been absorbed by the thirsty pores, then should follow warm water to modify and control the action of the oil. It will be seen at once that the oil has taken its place in the center of the leather, while the water is on the outside and may be evaporated easily; but if the water is inside it has to pass the oil, and there must be delay. If the oil is lodged in the center it must work toward the surface very slowly, and when evaporated, which will require months and even years, then it must be replenished in the same way.

It may be asked why soak with warm water at all, on this theory? Why not soak with oil and tallow exclusively? The answer is because the oil, under such circumstances, will "slough" off, but if water not exceeding 110 degrees in heat be applied it will drive in the oil from the surface and modify that greasy feeling and untidy appearance which attaches to over stuffed leather. Over stuffed leather can also be cured of its defects by being immersed in hot water to the extent above indicated. If no thermometer is at hand to test the degree of heat it will be safe to immerse the over stuffed leather for one minute in water as warm as the human hand and pulse can bear, and the result will be that all the excess of grease will be driven to the center, and the surface will dry fair. That no oil has passed out of the side will be demonstrated by the fact that no grease will appear on the surface of the water. It must then have gone in, which will be shown by the retained weight.

CHAPTER XXV.

DIRECTIONS FOR THE CONSTRUCTION OF DETACHED FURNACES FOR BURNING WET SPENT TAN.

BY THERON SKEEL, C. E.

The problem in designing a furnace and boilers for burning wet spent tan to furnish steam for a tannery is different from that in designing a furnace for coal or wood in that the fuel has generally no value for any other purpose, all that is not burned being thrown away. In case the tannery is not located where the spent tan can be run off through the bottom of the leach by a stream of water, but where it has to be shoveled from the leach and carted away, tanners will find it cheaper (so far as first cost) to burn all the tan they leach, for by so doing the boilers may be made smaller than would be necessary if they burned as little tan as they could get along with, but as the heat is more intense they will burn out sooner. A sole leather tannery can be run by burning about one-half the tan leached, if the ovens and boilers are well proportioned, while an upper leather tannery will need to burn nearly all the tan leached, even in the best form of furnaces and boilers; or, if the furnaces and boilers are of inferior design, will run short of bark at certain seasons of the year.

A tannery producing annually 20,000 sides of sole leather

will need for tanning about 2,000 cords of hemlock bark, measured before being ground, and *must* burn one-half of this, or 1,000 cords, to make all the steam used in the tannery to run engines and pumps, heat liquor, etc. The weight of an average cord of air-dried hemlock bark, as measured in the pile before being ground, will be 2,000 pounds. In the process of leaching this cord of bark will lose nearly 400 pounds, and the portion remaining will weigh 1,600 pounds in the same state of dryness as before being leached, but as it comes from the leaches will bring with it in addition 2,000 pounds of water, making the weight of the *wet spent tan* resulting from one cord of chip bark 3,600 pounds. The total weight of spent tan, therefore, produced each year by a 20,000-side sole leather tannery will be (3,600×2,000=) 7,200,000 pounds, and the weight of the portion that must be burned, being about one-half—3,600,000 pounds.

If the machinery of the tannery runs twelve hours a day and 300 days in the year, the total number of hours will be (300×12=) 3,600, and therefore the weight of wet spent tan that must be burned each hour in a furnace and boilers of the best construction and design (3,600,000÷3,600=) 1,000, or a thousand pounds an hour.

Table I is computed in this way.

TABLE I.—WEIGHT OF WET SPENT TAN THAT MUST BE BURNED EACH HOUR IN A FURNACE AND BOILER OF BEST CONSTRUCTION TO FURNISH ALL THE STEAM NECESSARY FOR THE TANNERY, IF THE MACHINERY RUNS 12 HOURS A DAY AND 300 DAYS A YEAR.

Number of Sides Tanned Annually.	Pounds of Wet Spent Tan Burned Each Hour.
10,000	500
20,000	1,000
50,000	2,500
80,000	4,000
100,000	5,000

Having determined how much tan must be burned, the next consideration is whether the oven shall be fed from the top or from the front. Except under very exceptional circumstances I consider it is better to feed from the *top*. If the oven is fed from the front one man cannot feed fast enough to keep up steam in a larger tannery than 60,000 sides, even if the tan is brought into the fire room floor by a laborer, (if he has to attend to the water in the boiler at the same time,) while, if the oven is fed from the top, one man can supply bark fast enough to keep up steam in a 200,000-side tannery. An oven fed from the top does not need nearly such close attention as an oven fed from the front. The former may be filled up and not touched again for an hour, if the fireman's attention is withdrawn, as often happens when the pump breaks down or in other cases, while the oven fed from the front must be supplied with bark every fifteen minutes. An oven fed from the top may be smaller than is admissible in an oven fed from the front, and requires less skill in firing. The only case when it would seem desirable to use an oven fed from the front is when it is very inconvenient to get the bark up on top of the oven, or when there is a small hight over the oven, as when the floor of the room above is so near the top of the oven as to be in danger from fire.

In order to successfully burn wet tan it is necessary that the surface of heated brick work surrounding the tan in the furnace should be large in proportion to the weight of tan burned, in order that this surface should not be cooled off by the water evaporated from the fresh charge of wet tan, below the temperature necessary to ignite the gases given off by that tan after it is dried and before it commences to burn. This condition requires a large oven and a slow rate of combustion. It is found in practice that the combustion is sensibly perfect when the wet spent tan is burned at the rate of 15 pounds on each square foot of grate surface per hour.

Table II is calculated from Table I in this way, and gives the necessary grate surface for a furnace fed from the top.

TABLE II.—AREA OF GRATE SURFACE NECESSARY IN A TANNERY BURNING ONE-HALF THE TAN LEACHED IN AN OVEN FED FROM THE TOP TO FURNISH ALL THE STEAM NEEDED, THE MACHINERY RUNNING 12 HOURS A DAY AND 300 DAYS IN THE YEAR.

Number of Sides Tanned Annually.	Area of Grate in Square Feet.
10,000	33
20,000	66
50,000	166
80,000	266
100,000	333

This amount of grate surface may be obtained in several ovens if necessary. The ovens had better be at least 6 feet wide, and as long as convenient. The feed holes should be two-thirds of the width of the oven between centers, that is, for an oven 6 feet wide the feed hole should be 4 feet between centers. The feed holes may be 12 to 18 inches in diameter at the top, and larger at the bottom, as the bark is fed in more easily through a large hole than a small one. The hight of the oven, measured from the grate bars to the crown, should be three-quarters of its width, that is, an oven 6 feet wide should be 4½ feet high from the crown to the grate.

In order to take up sufficient heat the heating surface in the boilers should be one-half of one square foot for each pound of wet spent tan burned per hour in an oven fed from the top. This amount of heating surface will reduce the temperature of the gas in the chimney to about 600°. Table III is computed from this proportion. [The heating surface is all the surface of the boiler, whether on the shell or in the flues or tubes, that has water on one side and the hot gas on the other.]

TABLE III.—AREA OF HEATING SURFACE NECESSARY FOR A TANNERY HAVING AN OVEN FED FROM THE TOP, AND BURNING ONE-HALF OF THE TAN LEACHED TO MAKE ALL STEAM USED, IF THE MACHINERY RUNS 12 HOURS A DAY AND 300 DAYS IN THE YEAR.

Number of Sides Tanned Annually.	Area of Heating Surface in Square Feet.
10,000	250
20,000	500
50,000	1,250
80,000	2,000
100,000	2,500

In case an oven fed from the front is used, the heating surface must be 20 per cent. more—that is, for a 10,000-side tannery, 300 square feet, etc.

In order to burn the wet spent tan at the rate given the chimney *must* be at least 70 feet high and must have an area of at least one-quarter of one square inch for each pound of wet spent tan burned per hour, but the performance will be more satisfactory if the chimney is made 100 feet high, and of an area of one-half of one square inch for each pound of wet spent tan burned per hour. The taller chimney will give a better draft, which will get up steam quicker after it has been allowed to run down, while the larger area will give a slower current in the chimney, and the sparks will not be so likely to be carried from the chimney and to endanger the adjacent buildings. The chimney may be made of sheet iron, but brick is recommended as a better material, or sheet iron lined with brick. The iron, if not lined with brick, in time rusts out, and cold air leaks in and spoils the draft. The brick when once built is permanent. A larger chimney than given in the following table (calculated at the rate of one-half of one square inch to each pound of wet spent tan burned per hour) is an advantage.

TABLE IV.—DIMENSION OF CHIMNEY NECESSARY TO CONSUME ONE-HALF OF TAN LEACHED WITHOUT ENDANGERING SURROUNDING BUILDINGS BY SPARKS CARRIED UP BY CURRENT OF GAS WHEN MACHINERY RUNS DURING 12 HOURS A DAY FOR 300 DAYS IN THE YEAR.

Number of Sides Tanned Annually.	Area in Square Inches.	Diameter of Chimney, if it is Circular, in Inches.	Sides of Chimney, if it is square, in Inches.
10,000	250	18	16
20,000	500	25	23
50,000	1,250	40	36
80,000	2,000	50	45
100,000	2,500	56	50

The size of the chimney may be the same whether the oven is fed from the top or from the front. There must be a damper in the chimney, or what is better, a damper between the outlet of each nest of boilers and the chimney, to regulate the draught.

The cross section of the flues or tubes in the boiler must be (in the case of an oven fed from the top) as great as the necessary area of chimney, or one-quarter of a square inch for each pound of wet spent tan burned per hour, for the gas is hotter and more bulky when it enters the flues than when it enters the chimney. Only one pound of wet spent tan can be burned for each one-quarter square inch of area of the flues or tubes in an oven fed from the front, but the flues or tubes may be made larger than this, and the draft checked by the dampers or ash-pit doors. Thus, if a tanner has an oven fed from the front and two horizontal flue boilers with two twelve-inch flues each, the total cross section is 452 square inches, and the greatest weight of wet spent tan that can be burned per hour is ($452 \div \frac{1}{4} =$) 1,808. But in order to get perfect combustion the flues should be twice as large as this, or one-half square inch area for each pound of wet spent tan burned per hour. It is of no advantage to have the flues or tubes of larger area than one-half of one square inch for each pound of wet spent tan burned per hour, (as given in table V,) and all tubes or flues added after that area is reached are *entirely useless.* The flues or tubes in an oven fed from the top may be made of any area from one-quarter to one-half of one square inch for each pound of wet spent tan burned per hour, and if the length of the boilers is proportioned to the area of the tubes so that the *heating surface is the same in each case* the effect will be the same, but short boilers with large tubes or flues are cheaper to build than longer ones with smaller flues or tubes.

TABLE V.—AREA OF FLUES OR TUBES THAT IS OF ADVANTAGE IN BURNING WET SPENT TAN IN ANY KIND OF OVEN, WHEN THE MACHINERY RUNS 12 HOURS A DAY AND 300 DAYS IN THE YEAR.

Number of Sides Tanned Annually.	Area of Tubes or Flues in Square Inches.
10,000	250
20,000	500
50,000	1,250
80,000	2,000
100,000	2,500

Horizontal tubular boilers are the cheapest, and if the water used is free from sediment or mineral water—that is, if the water comes from a clear stream of soft water—will last from eight to ten years, but as they contain but little water in comparison with the heating surface, must be fed often, and are likely to leak or to have the tubes burned out if the water is often allowed to get low and then quickly pumped up. If the water contains much sediment, or if it is "hard," the tubes are soon covered with a deposit of mud or scale. If the deposit is mud the boiler can be washed out, but if it is scale from "hard water" in a few years the tubes are coated so thick that they have to be taken out and new ones put in.

A tubular boiler of the dimensions given in the tables must

be fed every twenty minutes at least, and will occupy the unremitting attention of the fireman or engineer.

Plain cylinder boilers are the most expensive, but are the easiest kept in repair, and the easiest cleaned from mud or scale. They have, in addition, the disadvantage of requiring a larger boiler house and more expensive setting and foundations. They contain so much water that they may be neglected for several hours without danger, and the steam pressure will be more nearly constant than with horizontal tubular or flue boilers.

Horizontal flue boilers are more expensive than the tubular and less expensive than cylinder boilers, and they do not require so large a boiler house or so expensive a setting or foundation as the cylinder boilers, while they can be cleaned and repaired with more facility than the tubular boilers. They will last from sixteen to twenty years, or twice as long as the tubular. They contain more water than the tubular boilers, and (at the rate of combustion recommended in this paper) may be safely neglected for at least one hour without the water getting too low. If horizontal-flue or cylinder boilers are used, they may be connected together in nests of from two to six boilers, and each nest furnished with a feed, blow, safety and stop valve.

The cost of the boiler alone for a 20,000-side tannery will be in 1876:

1 horizontal tubular boiler, 54 inches diameter, 12 feet long, 43 3½-inch tubes, shell ⅜ inch. at 10c....... $700
Or, 3 horizontal-flue boilers, 40 inches diameter, 16 feet long, two 12-inch flues, each shell 5-16-inch, at 7½c.................................. $1,000
Or, 4 cylinder boilers, 30 inches diameter, 36 feet long, shell ¼ inch, at............................. $1,100

From this it appears that the flue and cylinder boilers cost nearly the same, but the flue boilers are cheaper to set. On the whole, flue boilers are to be recommended as the best unless the water is *very hard,* or unless it is particularly desirable to have a boiler that needs very little attention.

The best way to feed the boiler is by a pump worked from the main engine, and the next best way is by an injector. If the boiler is fed by an injector the heater to heat the feed water by the exhaust steam cannot be used, but this is not of importance, as the saving of 10 per cent. that may be made by the heater is of no moment when the fuel has no value, and the cost of the heater may be transferred to the boilers. If the boilers are fed by a steam or other pump the heater had better be used, as cold feed water has an injurious effect on the boilers. If they are fed by a pump worked by the main engine it is well to arrange the pump so that the stroke may be varied, and to adjust it to work *all the time.*

In case the tannery is not located where the spent tan can be thrown into the creek, or where there is any market for it, tanners will find it cheaper to burn it all rather than cart any away. In that case they must provide larger ovens, and will not need so large boilers as when they only burn one-half.

The wet spent tan may be burned in this wasteful manner (the object being to get rid of and not to economize it) at the rate of 30 pounds per square foot of grate per hour, and the ovens need to be made one-third larger than given in table, for same sized tannery, and the boilers may be made one-third *shorter,* but the diameter and number of tubes or flues must be the same, or the diameter had better be one-eighth more. The hight and diameter of chimney may be the same. That is, the ovens must be one-third longer and the boilers may be one-third shorter, if the tanner wishes to burn all the tan made, than if he burns only one-half. In this way the boilers are less expensive, but as they will wear out (owing to the higher temperature of the gas under them) in considerably less time, this arrangement of small boilers and large ovens is *not to be recommended.*

As an example in the application of the foregoing rules suppose it is desired to design a furnace to be fed from top, furnish steam for a tannery tanning 100,000 sides of sole leather, and running *night and day.* This would be equivalent to a 50,000 side tannery running 12 hours a day, and we find from the tables as follows:

Grate surface ....square feet.	$166\frac{2}{3}$
Heating surface...... ....square feet.	1,250
Cross section flues. ...square inches.	1,250
Diameter of chimney..........inches.	40
Pounds wet spent tan per hour...	2,500

The grate surface may be put in two ovens, each 6 feet wide and 14 feet long, with three feed holes, each 5 feet between centers, or may be put in one oven 7 feet wide and 24 feet long, with four feed holes, each 5 feet between centers. The last arrangement would be the cheaper, but the first is to be preferred, because it is well to have a pair of ovens, that one may be used if the other breaks down.

The boiler surface may be obtained in two horizontal tubular boilers, each 60 inches diameter and 12 feet long, with sixty $3\frac{1}{2}$ inch tubes, or in four horizontal flue boilers, each 42 inches in diameter and 22 feet long, with two 14 inch flues each; or eight cylinder boilers, each 30 inches in diameter and 36 feet long.

The chimney would be 40 inches in diameter, or 36 inches square. If it were desired to have an oven fed from the front, then the grate surface must be the same, and the heating surface 25 per cent. more on 1,600 square feet.

The depth of the grate must not be more than the fireman

can easily throw the tan, or from 6 to 7 feet, and therefore the width must be from 28 to 24 feet.

The grate surface could be obtained in two ovens, each 7 feet deep and 12 feet wide, or in three ovens, each 6 feet deep and 9 feet wide.

The boilers must have 25 per cent. more surface than for the oven fed from the top, so that the same boilers as were used for the furnace fed from the top will answer if they are made one-quarter longer, viz.:

Tubular boilers	15 feet long.
Flue boilers	28 feet long.
Cylinder boilers	45 feet long.

A still better arrangement would be six horizontal flue boilers, 40 inches in diameter with two 12-inch flues, 22 feet long.

In case the oven is fed from the front I consider that the grate bars had better be from one foot to one foot and a half below the sill of the doors through which the tan is fed, as it will be found easier to throw the tan to the back of the grate when they are so placed than when they are flush with the sill, as ordinarily arranged, and also for the same reason that they should be from six inches to a foot lower at the back than at the front, in place of being level as usual. The grate bars are placed flush with the door sill for a coal fire to facilitate cleaning the fire, which never need be done when tan bark is burned.

The furnace doors should be made with a register to be opened as the fire may require.

The ash pit doors should have registers, and the draft should be regulated by partly closing the registers in place of the damper.

One of the principal reasons why tan burning ovens often fail to give satisfaction is that the joints around the various openings and the sheet iron works are *poorly made*, and cold air leaks into the flues or chimney and chills the water or injures the draft. *No air should be admitted to mix with the gas after it has left the furnace.*

The space between the bottom of the grate bars and the floor of the ash pit should be at least two feet, and may be as much more as is convenient without detriment.

The space between the grate bars should not be more than ¼ inch, and may be as little as ⅛ inch. With this width of space (¼ inch) only a very insensibly small portion of the tan will fall through into the ash pit.

It has been customary in furnaces fed from the front to use the cone grate bars. I do not think these are any advantage, but that the furnace will work just as well with the ordinary bars, which are cheaper.

The grate bars, whether for an oven fed from the top or front, may be made much lighter than for a coal fire, as they have only to carry a load of about 30 pounds to the square foot in place of 60 pounds as in a coal fire, and are not broken by blows of the slide bar and hoe in cleaning fire. The heat they are called on to endure is only about two-thirds of a coal fire, and they may be safely made as light as 60 pounds per square foot when 31 inches long. When the tannery is so situated that it is convenient to burn *all the tan* the ovens may be made one-third larger and the boilers with one-third less heating surface. The chimney should also be one-third larger cross section, in order to avoid danger from sparks. Under these circumstances all the steam necessary will be made by the boilers, but all the tan made must be burned, and the gases will leave the boiler at nearly a red heat, or about 1,000°. This temperature will set fire to wood, and great care must be taken to keep the flue leading to the chimney out of contact with all woodwork.

Six plates illustrating the two kinds of furnaces described in this chapter are given in subsequent pages. The first three plates, I, II, III, are plans of a set of ovens to be fed from the top, which are located at Wilcox, Pa., and used to supply steam to the tannery at that place, which has a capacity of one hundred thousand hides per annum. These ovens and boilers are placed in a building distant three hundred feet from the nearest of the other buildings of the tannery.

Plate III, figure 1, shows a plan. In the left-hand portion of the figure the roof of the building is supposed to have been removed, so that the fire-room floor and the tops of the boilers are seen. In the right-hand portion the fire-room floor and the boilers are also supposed to have been removed, so that the grate bars and the dividing wall under the boilers are seen.

Figures 2 and 3, on the same plate, show east and west elevations. In figure 2 the rear wall of the right-hand oven is supposed to have been removed, so as to show the boilers and the dividing wall between. In figure 3 the front wall is supposed to have been removed so as to show the arches over the ovens, the grates and the fire-room floor.

Plate I is a section through one oven and boiler, showing the oven ash-pit grate, holes for feeding tan into oven, fire-room floor, the boilers, steam drum, mud drum and door to enter or sweep out flues in boiler. The two sections of wall marked "wall" in the plate are portions of the walls of the building, which continue up and support the roof.

Plate II, figure 2, shows the north elevation, in which the door through which the tan is brought from the leaches to the fire room is shown by the opening marked 8x10. Figure 1 on same plate shows two sections, one marked "Section on

(A B)" being through the oven, and the other marked "Section on (M N)" being through the boilers and brick work back of oven.

The furnace at Wilcox has been selected to illustrate this article in preference to a new design, because it is a furnace in actual use, which gives perfect satisfaction, and because, as it has been already built, the cost is accurately known. It will be seen that it really consists of two complete sets of ovens and boilers, the only thing in common between them being the chimney and fire room. Each pair of ovens has its own set of boilers, its own feed water tank marked "T" and "T'" in figure 1, plate III. Either pair of ovens would be nearly large enough to supply steam to run the whole tannery if the other pair were stopped, either by accident or for repairs. In case, however, only one pair of ovens were used, it would probably be necessary to burn nearly *all the bark leached*, while, when both pairs are used, it is only necessary to burn about one-half the bark. Each pair of these ovens is probably capable (if forced to the utmost) of burning one and a half cords of wet spent tan per hour, which would evaporate 9,000 pounds of steam, being equivalent to about 180 horse power. In actual practice at Wilcox they burn one-half a cord of wet spent tan each, each pair of ovens per hour evaporating 8,000 pounds of water, being equivalent to 160 horse power from both ovens. This steam runs the whole tannery, except the rolling loft.

The foundations of the ovens and chimney are not shown in the plans. They should be of stone and should go down in the ground below the action of the frost. The foundation under the chimney, on account of the great weight it has to bear, should go deeper than this, or at least six feet below the original surface of the ground, and should be a foot larger than the base of the chimney all around. The first step in the construction of this furnace would be to mark out on the ground an area two feet larger each way than the outline of the building, and then excavate all the earth from this space three feet deep. Next, on the bottom of this excavation, mark out the position of the walls, and dig a trench two and a half feet wide and two and a half feet deep under each. Thus there would be four trenches under the four walls of the building, and in addition, four trenches parallel to the side walls under the four dividing walls of the ovens and boilers. The foundation for the chimney will come in the center, and should be fourteen feet square and go down four feet below the bottom of the excavation, or seven feet below the original surface of the ground. The foundations under the two wings containing the feed-water tanks may be two feet wide, and sunk three feet below the original surface of the ground. The walls of the ovens are commenced on these foundations, three and a half feet thick, but with a batir of one foot in the hight from the foundation up to the level of the fire room floor. From this point they are perpendicular, and of a uniform thickness of sixteen inches. At the fire room floor, therefore, the oven walls are two and a half feet thick. The walls of the feed-water tank rooms may be sixteen inches thick.

The dividing wall between the ovens has a uniform thickness of one foot ten inches up to the level of the grate bars. The fire brick linings of the ovens commence at the level of the grates, and are set back two inches on each side to leave a shoulder for the grate bars to rest on. The grate bars are each three feet long. One end rests on this shoulder, and the other on a wrought iron bar (a worn out piece of railroad track answers very well), which runs the whole length of the furnace, and is embedded on the walls at each end. There are four one and one-quarter inch iron rods, each twenty feet long, with a thread and nut on each end, and cast iron washers, through each oven, under this bearing bar, to tie the side walls together. The whole inside of the ovens, and the tops of the bridge walls, and the inside of the walls under the boilers, are lined with fire brick. The holes for feeding the tan through the top of the arch into the oven are lined with fire brick which must be made of the required shape by the manufacturer. The side walls of the ovens rise vertically for one foot, and are then turned into a circular arch of three feet radius. The lining of this arch should be of fire brick not less than six inches thick, and there should be in addition two courses of common brick, so that the whole thickness of the arch is fourteen inches. The fire room floor may be of brick, laid on edge, and the space between the arches under the floor filled up with broken bricks and mortar. There are four doors through each outside wall of the ovens. These must be made with arched tops of fire brick, made the exact shape by the manufacturer.

The owner of the furnace will find it to his advantage to have the brickwork around the ovens constructed with great care, as if carelessly built it will crack, or even fall down, in a few years. The heat in one of these tan-burning ovens, although not so great as in a coal fire, is greater than in the ordinary furnace when the grate to burn the coal is *under the boiler*. A plan of the grate bar to be used is shown in figure 3, plate V. These grate bars may be much lighter than when used for coal. The oven doors should each have a register in them, to be partly opened to admit air above the fire when the oven is in use.

Through the courtesy of Mr. Judson Schultz, of Wilcox, Pa., I am enabled to exhibit the exact cost of this furnace in 1872. At the present time (September, 1876) the cost would probably be 25 per cent. less.

COST OF FURNACE AT WILCOX, 1872.

Brickwork—		
Chimney—35,000 bricks, at $6 00	$210 00	
Two masons, 20 days, at $5 00 per day	200 00	
Three helpers, 20 days, at $1 50	90 00—	$500 00
Ovens and Walls—112,500 common brick, at $6 00	675 00	
5,800 fire brick, at $60 00	348 00	
One-quarter-circle bricks, for arch	100 00	
Three masons, 52 days each, at $5 00 per day	780 00	
Two helpers, 50 days each, at $1 50 per day	150 00	
Digging foundations, hauling stone and sand	300 00—	2,353 00
Boilers Etc.—		
Six boilers and standards, with two mud drums		3,575 57
Four furnace doors, four back connection doors, four doors to enter under boiler, two smoke jacket doors, one chimney door, 5,536 lbs, at 5½c	303 48	
Back stays, 8,062 lbs, at 5c	403 10	
2,048 grate bars, 26,175 lbs, at 4½c	1,177 88	
Washers, 130 lbs, at 5½c	7 15	
Dampers and smoke jacket	58 00	
Tie rods under grate	100 00—	2,049 61
Roof—		
3,000 feet of lumber for rafters, at $10 00	30 00	
Labor	50 00	
Sheet iron	325 00—	405 00
Total cost of oven and fixtures		$8,883 18

Taking the total cost at $7,000, it will appear that the original investment in the furnaces will be at the rate of 7 cents for each hide tanned annually, and that at 15 per cent. (allowing 7 per cent. for interest on capital, 3 per cent. for annual repairs, and 5 per cent. for a sinking fund), the annual cost will be $7,000×.15=$1,050, or about *one cent for each hide tanned.*

If it were required to build a cheaper furnace than that at Wilcox, the brick walls above the ovens and boilers and the iron roof may be omitted, and wooden walls and a wooden roof used in their stead.

If tubular boilers in place of flue boilers are used, the cost of the boilers will be less, and the brickwork setting will cost less.

If the feed water is free from sediment and scale, tubular boilers will be just as economical as flue boilers, but will not last so long, and will need more unremitting attention on the part of the engineer to keep the water at the correct level. On the whole, they are not to be recommended.

Plates IV, V, VI, represent an oven for burning wet spent tan when the tan is shoveled in to the oven through an ordinary furnace door like coal or wood.

Plate V is a section through the center of oven and middle boiler, showing the oven grate and ash pit, the boiler, steam and mud drum, and the front and back connections.

Plate VI, figure 1, represents, on the left hand portion, an elevation of the front of the oven, showing one furnace and one ash pit door and half of the middle furnace door, and on the right hand portion a section through the oven showing the top and sides of the oven, the thickness of the brickwork and the grate bars. Figure 2 is a section through the boiler, back of the oven, showing the three boilers and flues, the steam and mud drum and the thickness of the brickwork. The unshaded bricks represent the fire brick in all three plates. Furnaces fed from the front like this are generally considered cheaper to build than those fed from the top, but this is because they are generally made smaller in proportion to the size of the tannery than those fed from the top, and are therefore ordinarily much less efficient.

If both furnaces are built of the size required to give the best results, the furnace fed from the front will require to be the largest and will cost the most.

The construction will be readily understood from the plans. The plan of the grate shown in figure 3, plate V, has been found in practice to be strong enough. There are two lengths of these grate bars, each 2 feet 8½ inches long. The back length is level, but the front length should be raised from six to eight inches in front to facilitate feeding the tan toward the back. The ends of the grate bars must be carried by

16

wrought iron bars, reaching across the oven and built into the brick work at each side. An old piece of railroad iron will answer, or a piece of wrought iron four inches deep and one inch thick.

The furnace and ash pit doors may be made of a piece of boiler iron a quarter of an inch thick, with the hinges and latch riveted on. The lugs to carry the hinges, and the latch, may be built into the brick wall, thus dispensing with the cast iron door frames. The furnace door should be double, the inner lining of boiler plate being punched full of quarter inch holes two inches apart. Both the ash pit and the furnace doors should have openings through them with registers.

The three furnace doors should be each 20 inches wide and 18 inches high on the outside, spreading out wider on the inside, as shown in the plan, plate IV. The ash pit doors should be 16 inches wide by 24 inches high in the clear. The fire room floor should be 24 inches below the sill of the furnace door, or 18 inches above the bottom of the ash pit, as the fireman can shovel in the tan easier than if the floor is as low as the bottom of the ash pit.

The brick work around the oven must be tied up with buck stays and iron rods, as shown in the plans, for as the arch over the top of the oven has a wide spread, it exerts a great pressure on the side walls. Also the fire bricks in the arch must be laid with care with narrow joints, or they will fall out after the furnace has been a little while in use.

The smoke connection in this furnace is shown in plate V, as built of brick running across the boilers over the top of the oven. This connection must lead to the chimney built on either side. The chimney should be eighty feet high and twenty-four inches in diameter. The back ends of the boiler are carried by the mud drums, and the front ends by a piece of railroad iron built into the brick work over the oven.

TABLE SHOWING THE COST OF LEATHER MADE FROM GREEN HIDES, INCLUDING COMMISSIONS, INTEREST, ETC.,

With the Price of Hides Ranging from 5 to 15 cents per Pound, and the "Gains" from 40 to 100.

BY THEODORE SCHULTZ.

Gain. Price	40	45	50	55	60	65	70	75	80	85	90	95	100
5 ..	26 61	25 08	23 86	22 86	22 03	21 32	21 73	20 19	19 74	19 33	18 97	18 65	18 36
5¼ ..	27 30	25 69	24 41	23 36	22 49	21 64	21 11	20 56	20 08	19 65	19 28	18 94	18 64
5½ ..	27 99	26 30	24 96	23 86	22 74	22 17	21 50	20 93	20 42	19 98	19 58	19 23	18 91
5¾ ..	28 67	26 91	25 51	24 36	23 40	22 59	21 90	21 29	20 77	20 30	19 89	19 52	19 19
6 ..	29 36	27 53	26 06	24 86	23 86	23 01	22 29	21 66	21 11	20 62	20 19	19 81	19 46
6¼ ..	30 05	28 14	26 61	25 36	24 32	23 44	22 68	22 03	21 43	20 97	20 50	20 10	19 74
6½ ..	30 74	28 75	27 16	25 86	24 78	23 86	23 07	22 39	21 80	21 27	20 80	20 39	20 01
6¾ ..	31 42	29 36	27 71	26 36	25 23	24 28	23 47	22 76	22 14	21 60	21 11	2 68	20 29
7 ..	32 11	29 97	28 26	26 86	25 69	24 70	23 86	23 13	22 49	21 92	21 41	20 97	2 56
7¼ ..	32 80	30 58	28 81	27 36	26 15	25 13	24 25	23 49	22 83	22 24	21 72	21 25	20 84
7½ ..	33 49	31 19	29 36	27 86	26 61	25 55	24 65	23 86	23 17	22 57	22 03	21 54	21 11
7¾ ..	34 17	31 80	29 91	28 36	27 07	25 97	25 04	24 23	23 52	22 89	2 34	21 83	21 39
8 ..	34 86	32 41	30 46	28 86	27 53	26 40	25 43	24 59	23 86	23 21	22 64	22 12	21 66
8¼ ..	35 55	33 02	31 01	29 36	27 98	26 82	25 82	24 96	24 20	23 54	22 95	22 41	21 94
8½ ..	36 24	33 64	31 56	29 86	28 44	27 24	26 22	25 33	24 55	23 86	23 25	22 70	22 21
8¾ ..	36 92	34 25	32 11	30 36	28 90	27 67	26 61	25 69	24 89	24 18	23 56	22 99	22 49
9 ..	37 61	34 86	32 66	30 86	29 36	28 09	27 00	26 06	25 24	24 51	23 86	23 28	22 76
9¼ ..	38 30	35 47	33 21	31 36	29 82	28 51	27 40	26 43	25 58	24 83	24 17	23 57	23 04
9½ ..	38 99	36 09	33 76	31 86	30 28	28 93	27 79	26 79	25 92	25 15	24 47	23 86	23 31
9¾ ..	39 67	36 70	34 31	32 36	30 74	29 36	28 18	27 16	26 27	25 48	24 78	24 15	23 59
10 ..	40 36	37 31	34 86	32 86	31 19	29 78	28 57	27 53	26 61	25 80	25 08	24 44	23 86
10¼ ..	41 05	37 92	35 41	33 36	31 65	30 20	28 97	27 89	26 96	26 12	25 39	24 73	24 14
10½ ..	41 74	38 53	35 96	33 86	32 11	30 63	29 36	28 26	27 30	26 45	25 69	25 02	24 41
10¾ ..	42 42	39 14	36 51	34 36	32 57	31 05	29 75	28 63	27 64	26 77	26 00	25 31	24 69
11 ..	43 11	39 75	37 06	34 86	33 03	31 47	30 15	28 99	27 99	27 09	26 30	25 60	24 96
11¼ ..	43 80	40 36	37 61	35 36	33 48	31 90	30 54	29 36	28 33	27 42	26 61	25 89	25 24
11½ ..	44 49	4 97	38 16	35 86	33 94	32 32	30 93	29 73	28 67	27 74	26 92	26 18	25 51
11¾ ..	45 17	41 58	38 71	36 36	34 40	32 74	31 32	30 09	29 02	28 07	27 22	26 47	25 79
12 ..	45 86	42 20	39 26	36 86	34 86	33 17	31 72	30 46	29 36	28 39	27 53	26 76	26 06
12¼ ..	46 55	42 81	39 81	37 36	35 32	33 59	32 11	30 83	29 70	28 71	27 84	27 04	26 34
12½ ..	47 24	43 42	40 36	37 86	35 78	34 01	32 50	31 19	30 05	29 04	28 14	27 33	26 61
12¾ ..	47 92	44 03	40 91	38 36	36 23	34 44	32 90	31 56	30 39	29 36	28 45	27 62	26 89
13 ..	48 61	44 64	41 46	38 86	36 69	34 86	33 29	31 93	30 74	29 68	28 75	27 91	27 16
13¼ ..	49 30	45 25	42 01	39 36	37 15	35 28	33 68	32 29	31 08	30 01	29 06	28 20	27 44
13½ ..	49 99	45 86	42 56	39 86	37 61	35 71	34 07	32 66	31 42	30 33	29 36	28 49	27 71
13¾ ..	50 67	46 47	43 11	40 36	38 07	36 13	34 47	33 03	31 77	30 65	29 67	28 8	27 99
14 ..	51 36	47 08	43 66	40 86	38 53	36 55	34 86	33 39	32 11	30 98	29 97	29 07	28 26
14¼ ..	52 05	47 69	44 21	41 36	38 98	36 97	35 25	33 76	32 45	31 30	30 24	29 36	28 54
14½ ..	52 74	48 30	44 76	41 86	39 44	37 40	35 65	34 13	32 80	31 61	30 59	29 65	28 81
14¾ ..	53 42	48 92	45 31	42 36	39 90	37 82	36 04	34 49	33 14	31 95	30 89	29 94	29 09
15 ..	54 11	49 53	45 86	42 86	40 36	38 24	36 43	34 86	33 49	32 27	31 19	30 23	29 36

TABLE SHOWING THE COST OF LEATHER MADE FROM DRY HIDES, INCLUDING COMMISSIONS, INTEREST, ETC.,

With the Price of Hides Ranging from 15 to 30 cents per Pound, and the "Gains" from 140 to 200.

BY THEODORE SCHULTZ.

Gain. Price	140	145	150	155	160	165	170	175	180	185	190	195	200
15 ..	22 35	21 94	21 56	21 21	20 87	20 56	20 27	19 99	19 73	19 48	19 24	19 02	18 81
15½ ..	22 74	22 32	21 93	21 56	21 22	20 89	2 79	20 30	20 63	19 78	19 53	19 30	19 09
16 ..	23 13	22 70	22 29	21 91	21 56	21 22	20 91	20 62	20 34	20 07	19 82	19 59	19 36
16½ ..	23 52	23 07	22 60	22 27	21 90	21 56	21 24	20 93	20 64	20 37	20 11	19 87	19 64
17 ..	23 92	23 46	23 03	22 62	22 25	21 89	21 56	21 25	20 95	20 67	20 40	20 15	19 91
17½ ..	24 31	23 84	23 39	22 98	22 59	22 23	21 88	21 56	21 25	20 97	20 69	20 43	20 19
18 ..	24 70	24 22	23 76	23 33	22 94	22 56	22 20	21 87	21 56	21 26	20 98	20 71	20 46
18½ ..	25 10	24 59	24 13	23 69	23 28	22 89	22 53	22 19	21 87	21 56	21 27	21 00	20 74
19 ..	25 49	24 97	24 49	24 04	23 62	23 23	22 85	22 50	22 17	21 86	21 56	21 28	21 01
19½ ..	25 88	25 35	24 86	24 40	23 97	23 56	23 18	22 82	22 48	22 14	20 85	21 56	21 29
20 ..	26 27	25 73	25 23	24 75	24 31	23 89	23 50	23 13	22 78	22 45	22 14	21 84	21 56
20½ ..	26 67	26 11	25 59	25 11	24 65	24 23	23 82	23 45	23 09	22 75	22 43	22 12	21 84
21 ..	27 06	26 49	25 96	25 46	25 00	24 56	24 15	23 76	23 39	23 05	22 72	22 41	22 11
21½ ..	27 45	26 87	26 33	25 82	25 34	24 89	24 47	24 07	23 70	23 34	23 01	22 69	22 39
22 ..	27 85	27 25	26 69	26 17	25 68	25 23	24 80	24 39	24 00	23 64	23 30	22 97	22 66
22½ ..	28 24	27 63	27 06	26 53	26 03	25 56	25 12	24 70	24 31	23 94	23 59	23 25	22 94
23 ..	28 63	28 01	27 43	26 88	26 37	25 89	25 44	25 02	24 61	24 24	23 88	23 53	23 21
23½ ..	29 02	28 39	27 79	27 24	26 71	26 23	25 77	25 33	24 92	24 53	24 46	23 82	23 49
24 ..	29 42	28 77	28 16	27 59	27 06	26 56	26 09	25 55	25 23	24 83	24 45	24 10	23 76
24½ ..	29 81	29 15	28 53	27 95	27 40	26 89	26 41	25 96	25 53	25 13	24 74	24 38	24 04
25 ..	30 20	29 53	28 89	28 30	27 75	27 23	26 74	26 27	25 83	25 42	25 03	24 66	24 31
25½ ..	30 60	29 90	29 26	28 66	28 09	27 56	27 06	26 59	26 14	25 72	25 32	24 94	24 59
26 ..	30 99	30 28	29 63	29 01	28 44	27 89	27 38	26 90	26 44	26 02	25 61	25 23	24 86
26½ ..	31 38	30 66	29 99	29 36	28 78	28 23	27 71	27 22	26 75	26 32	25 90	25 51	25 14
27 ..	31 77	31 04	30 36	29 72	29 12	28 56	28 03	27 53	27 06	26 61	26 19	25 79	25 41
27½ ..	32 17	31 42	30 73	30 08	29 46	28 89	28 35	27 85	27 37	26 91	26 48	26 07	25 69
28 ..	32 56	31 80	31 09	30 43	29 81	29 23	28 68	28 16	27 67	27 21	26 77	26 35	25 96
28½ ..	32 95	32 18	31 46	30 79	30 15	29 56	29 00	28 47	27 98	27 51	27 06	26 64	26 24
29 ..	33 35	32 56	31 83	31 14	30 50	29 89	29 32	28 79	28 28	27 80	27 35	26 92	26 51
29½ ..	33 74	32 92	32 19	31 50	30 84	30 23	29 65	29 10	28 58	28 10	27 64	27 20	26 79
30 ..	34 13	33 32	32 66	31 85	31 19	30 56	29 97	29 42	28 89	28 40	27 93	27 48	27 06

The above tables represent the cost of leather made from green and dry hides at given prices of hides, tanning and commissions. The cost of leather from green hides includes 11 cents per pound for tanning, 5 per cent. commission for buying the hides. 5 per cent. for interest on money invested in hides (which is equivalent to interest for about eight and a half months), and 6 per cent. commission on an average price of 31 cents per pound for leather. On dry hides 9 cents per pound is allowed for tanning; commission on hides and interest are estimated the same as on green hides, and the commission on leather is calculated on an average price of 26 cents per pound.

LEATHER INDUSTRY OF THE UNITED STATES,

Including Leather Tanned, Leather Curried, Enameled and Patent Leather, Morocco and Dressed Skins.

[COMPILED FROM THE UNITED STATES CENSUS FOR 1870.]

States and Territories.	Establishments.	Steam Engines.		Water Wheels.		Hands Employed.				Capital.	Wages.	Materials.	Products.
		Horse Power.	Number.	Horse Power.	Number.	All.	Males above Sixteen.	Females above Fifteen.	Youths.				
Alabama	141	18	2	50	3	278	262	7	9	$206,769	$48,628	$263,744	$412,335
Arkansas	35	....	....	....	....	53	51	..	1	32,100	5,260	52,777	83,021
California	70	149	10	2	1	262	257	3	2	433,540	111,764	838,164	1,174,924
Connecticut	75	111	8	299	18	366	332	4	..	592,650	157,747	990,120	1,317,030
Delaware	28	267	13	....	....	609	522	72	15	930,018	318,035	1,451,022	2,050,846
District of Columbia	3	63	3	....	....	32	32	..	..	85,000	10,700	69,272	146,475
Florida	3	....	....	....	....	11	11	..	..	8,000	3,600	8,040	13,800
Georgia	186	72	5	99	10	316	310	1	5	186,247	47,758	418,793	572,306
Illinois	98	568	26	5	1	754	751	1	2	1,286,300	372,944	3,241,197	4,150,338
Indiana	353	682	41	5	1	838	818	2	13	1,179,550	227,450	1,815,125	2,461,549
Iowa	35	3	1	....	....	55	55	..	..	38,175	6,825	70,823	94,449
Kansas	6	....	....	....	....	11	11	..	..	6,300	2,000	22,427	26,427
Kentucky	182	227	15	1	1	448	436	1	11	724,340	118,816	1,297,997	1,093,574
Louisiana	10	....	....	....	....	20	20	..	..	16,550	3,450	17,980	32,450
Maine	200	661	30	1,439	93	1,020	1,008	1	11	1,864,949	360,126	3,956,839	4,911,781
Maryland	125	286	23	35	3	547	519	12	16	1,120,125	195,911	1,516,861	2,084,696
Massachusetts	386	3,326	202	619	50	5,553	5,478	32	43	7,577,926	3,152,399	26,106,013	33,457,975
Michigan	173	938	53	194	9	734	729	2	3	1,307,540	285,749	2,013,156	2,689,408
Minnesota	12	8	2	....	....	30	30	..	..	34,400	11,150	79,747	107,007
Mississippi	56	....	....	14	3	101	99	..	2	49,660	13,293	184,367	262,723
Missouri	80	115	4	....	....	240	238	..	2	189,250	113,825	530,234	834,948
New Hampshire	126	405	22	1,245	60	667	665	..	2	1,223,600	271,638	3,037,894	3,744,236
New Jersey	148	701	38	150	10	1,298	1,239	33	26	2,679,487	839,038	7,529,916	9,307,948
New York	1,002	5,844	264	7,707	397	8,109	7,957	38	114	16,147,578	3,660,855	26,573,601	36,569,600
North Carolina	222	58	5	66	8	308	302	..	6	159,828	31,830	265,310	364,308
Ohio	892	1,853	115	94	7	2,115	2,076	3	36	3,304,841	652,919	5,755,256	7,332,982
Oregon	24	10	1	22	4	59	59	..	..	47,400	18,700	98,393	147,243
Pennsylvania	1,495	5,711	317	2,184	133	6,880	6,654	127	99	15,317,765	2,587,699	20,793,080	28,899,406
Rhode Island	14	135	5	27	2	213	213	..	..	546,000	108,455	1,467,258	1,828,264
South Carolina	65	....	....	70	8	127	123	2	2	40,200	13,386	1 6,225	166,025
Tennessee	396	310	12	68	8	769	753	2	14	705,665	152,775	1,401,765	1,851,638
Texas	56	....	....	....	....	90	89	..	1	54,843	11,935	77,068	117,911
Utah	37	....	....	21	3	56	53	..	3	45,300	11,405	59,353	91,603
Vermont	155	256	14	1,163	82	475	471	4	..	837,133	166,995	1,609,857	2,052,913
Virginia	318	86	7	105	7	507	500	..	7	302,484	62,799	591,090	790,443
Washington	10	....	....	20	2	16	16	..	..	17,000	4,700	36,161	51,399
West Virginia	178	222	15	31	4	328	319	1	8	456,379	68,892	626,678	640,245
Wisconsin	174	682	43	41	6	984	965	5	14	1,369,740	375,013	3,575,061	4,543,061
Total	7,569	23,407	1,296	15,776	940	35,243	34,423	353	467	$61,124,812	$14,506,775	$118,589,034	$157,237,097

STATISTICS OF THE SHOE AND LEATHER TRADES, AND COLLATERAL BRANCHES.

[COMPILED FROM THE UNITED STATES CENSUS FOR 1870.]

	Establishments	Steam Engines.		Water Wheels.		Hands Employed.				Capital	Wages	Materials	Products
		Horse Power	Number	Horse Power	Number	All	Males above 16	Females above 15	Youths				
* Boots and shoes	23,428	2,9 2	267	167	24	135,889	113,415	19,113	3,361	$4[illegible],094,516	$51,972,712	$93,582,528	$181,644,090
† Boots and shoes	3 151	....	....	....	..	91,702	70,668	18,208	2,806	37,519,019	41,504,444	80,5[illegible]2,718	146,704,085
Boot and shoe findings	271	310	32	223	14	2,773	1,045	1,442	286	358,500	792,857	1,817,[illegible]28	3,380,091
Leather, tanned	4,237	19,572	1,045	14,202	855	20,784	20,423	98	263	42,720,505	7,934,416	63,069,491	86,170,[illegible]83
Leather, curried	3,063	2,992	174	887	56	10,027	9,907	57	63	12,303,785	4,154,114	43,565,593	54,191,167
Leather, morocco	113	663	48	16	3	3,006	2,740	182	84	3,854,072	1,678,226	6,623,066	9,997,460
Leather, patent and enameled	26	354	14	45	1	528	509	....	19	906,000	341,445	3,211,749	4,018,115
Leather, dressed skins	110	206	15	626	26	598	844	16	38	1,340,450	397,574	2 099,735	2,859,972
Leather, belting and hose	91	302	13	42	3	608	784	8	16	2,118,577	454,187	3,231,204	4,558,043
Saddlery and harness	7,607	172	12	43	3	23,557	22,716	375	466	13,935,961	7 046,207	16,068,310	32,709,981
Trunks, valises and satchels	222	358	15	55	4	3,479	2,798	457	224	2,185,964	1,810,798	3,315,038	7,725 488
Leather board	8	80	4	495	13	94	87	5	2	268,000	38,260	135,675	242.500
Shoe pegs	26	257	10	305	18	279	175	28	6	169,900	78 051	63,736	264,847
Sumac ground	19	208	10	95	6	85	84	1	....	167,400	31,325	164,702	267,180
Bark ground	33	513	13	169	14	133	131	....	2	322,700	47,069	194,491	372,829
Furs, dressed	182	76	6	10	1	2,903	1,306	1,525	72	3,472.267	1,042,305	4,816,122	8,903,052

* All who make goods.

† Establishments making goods to the amount of $5,000 or more per year.

ILLUSTRATIONS.

Until within a few years it has been considered indispensable that the "sweat pits" of the tanner should be under ground, but now they are placed on the top of the ground, with earth or some other non-conductor surrounding and covering them. It is far more desirable that the bottom of the pit should be on a level with the beam house floor, or so nearly so that a wheelbarrow can be run from one to the other, than that there should be an earth or rock excavation surrounding the pit as was formerly thought necessary. The present plan enables the pits to be properly lighted both from the ends and tops.

These pits should be large enough to properly hang one pack each, whatever may be the size of that pack—usually from 100 to 200 sides. The ceiling should be at least eight, and if ten feet high all the better. There should be a space of at least two feet from the lower portion of the sides, while hanging, to the floor, for the purpose of allowing the free introduction of steam under the pack, if the pit is too cold, and of cold spring or well water if too warm. The temperature should be about 60 to 70 degrees. The air should never be dry, but kept moist by the introduction of steam or cold water. Usually the atmosphere will regulate itself. The wet pack, hung in an atmosphere of 65 degrees, will impart its moisture, which will condense so that drops of cold water will stand on the surface and show on the hair of the hide, precisely as from the ox when profusely sweating from over exertion.

The mode of hanging will vary with the judgment of the tanner. Some hang over sticks, others hang from tenter hooks. But whatever way hung, the sides or hides should be so placed that the attendant can readily reach each side, so that as soon as the hair of even one side or hide begins to "come" it should be dropped to the bottom, and when that of a majority of the pack "starts" they should be re-

VIEW IN PERSPECTIVE—SHOWING INSIDE OF TWO END PITS.

INTERIOR VIEW OF SWEAT PIT.

moved; but if proper care has been taken to have the whole pack in the same condition and of the same weight and substance the whole pack will come out about the same time. From five to seven days' time will sweat an ordinary dry hide in one of these pits. The ventilation of these pits should always be under the control of the attendant.

One of the views presented gives a lengthwise view of several pits, and the other that of a single pit as seen from the door.

THE "SHOVER" HIDE MILL.

The three sectional plates herewith given present a fairly intelligent working plan of the ordinary "fulling stocks" for softening hides. This form of mill has so completely taken the place of the old-fashioned "falling stocks" as to render it unnecessary to present a drawing of that discarded machine. The capacity or power of this mill to soften dry hides is almost without limit. One mill of the size here contemplated, run at the rate of sixty to eighty revolutions per minute, will soften 1,000 hides of ordinary weight in one week, running for twelve hours per day. It is within the writer's knowledge that a mill of this description has been made to soften 100,000 dry hides in one year, being run night and day.

In the drawing this mill is geared from above, but the power is often attached from beneath, and where this can be done it has the advantage of greatly facilitating the workmen in feeding the mill and handling the hides. With the latter attachment the arms of the mill are extended down through the hammers, and the bottom of the mill and driving shafts are attached below, just as they are represented in the drawing to be attached above. These mills, as formerly made, of hard wood plank, would wear out with one year's constant service; but lined and faced with cast and wrought iron, as

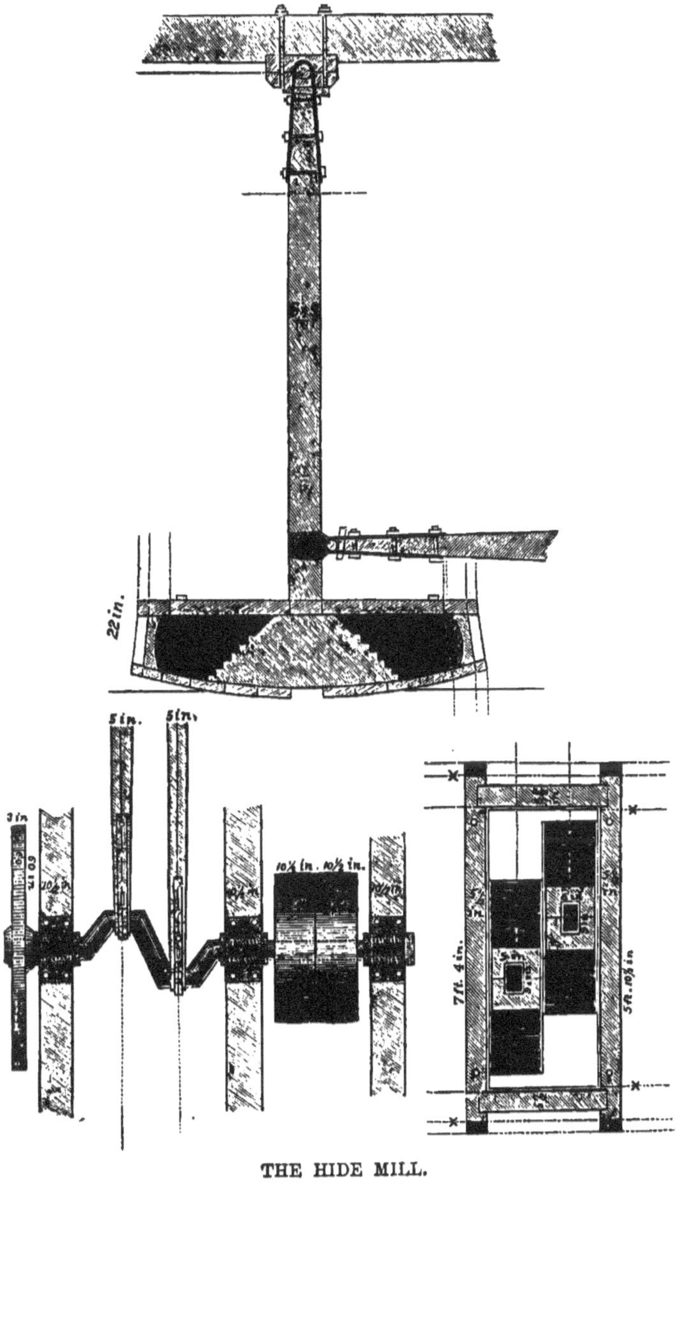

THE HIDE MILL.

is now customary, they will last for many years. The form of the mills varies slightly with the different makers, and the size as well. Tanners who are about to adopt this method of softening hides will do well to examine with care the most approved *angles* and *motion* of mills now in use in our best yards. The double action of these mills is as well adapted for skins and pelts of all kinds as for dry hides. There can be no doubt that from this form of mill the idea of the modern "washing machine" was taken, and in turn the tanner is indebted to the clothiers, and chamois and oil dressing leather manufacturers for the ideas contained in this most efficient hide softener, for with some variations of form it is a reproduction of the fulling stocks of the clothier a century ago. It was introduced among American tanners about the year 1830, at Salem, Mass., and did not find general acceptance among hemlock tanners in the State of New York until 1850.

THE HAND REEL.

On the opposite page will be found a representation of a hand reel, which is commended in Chapter VII., on "handling." The facility with which packs may be thus transferred from one vat to another commends this skeleton reel to all tanners. It is safe to estimate the performance of this machine with two men as equal to that of six men by the old hand process. Besides, it does not require either man to stoop in his work, and the labor is therefore much easier. The stand and skeleton drum should be made of as light material as possible, so that its transfer from one vat to another may be effected by the two men with ease. As there need be but one of these reels in any ordinary sized tannery, the tanner can well afford to have the frame, drum and bearings made of substantial but light materials, well adjusted in

THE HAND REEL.

all parts, even with brass bearings. The whole need not cost over ten dollars.

If still more economy of labor is required, these reels may be driven by power from shafting and pullies permanently running overhead. But it will be found that two men can shift ten thousand sides in ten hours without the use of power.

The sides may be tied together with strings or connected with a tie-loop; strings are preferable, and are quite inexpensive.

THE ROCKER HANDLER.

The accompanying drawing will serve to give a correct idea of the rocker handler, which is now held in such high esteem.

The gentle and intermittent motion required should be given from shafting from above. This shafting should run over the center aisle, and have projecting arms, from which there should extend a connection to each of the frames of the rockers. This connection may be by a pole or strip of plank two inches square, made of any tough timber, and so adjusted as to be readily disconnected. The vibration of the rocker on which the leather is hung should not be over four or six inches, thus causing as little agitation of the liquor as is consistent with a gentle movement of the fiber of the green stock. (See Chapters VII. and VIII.)

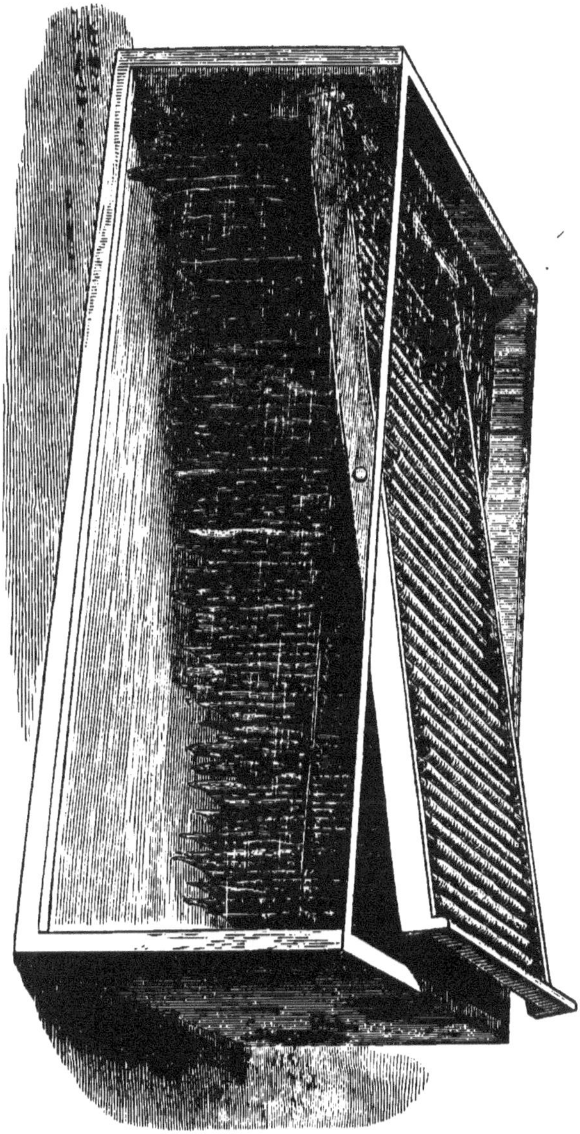

THE ROCKER HANDLER.

THE AMERICAN LEATHER ROLLER.

The skeleton drawing on the opposite page represents the universal American roller. The only patent now existing on this roller, as here presented, is on the leverage under the table. The whole cost of this leverage is $30, on which there may be a profit of $10 for the patentee. The roller bed is made of wood or metal; the one in the drawing is made of copper. It is cast hollow, and is valued at about $40. The whole machine will cost about $200, including the woodwork, table, etc. Such a machine will roll about 100 sides each day of ten hours.

The amount of pressure which can be brought to bear on the side has absolutely no limit. It is claimed that a side of sole leather can be cut in two if the whole power of the lever is evoked.

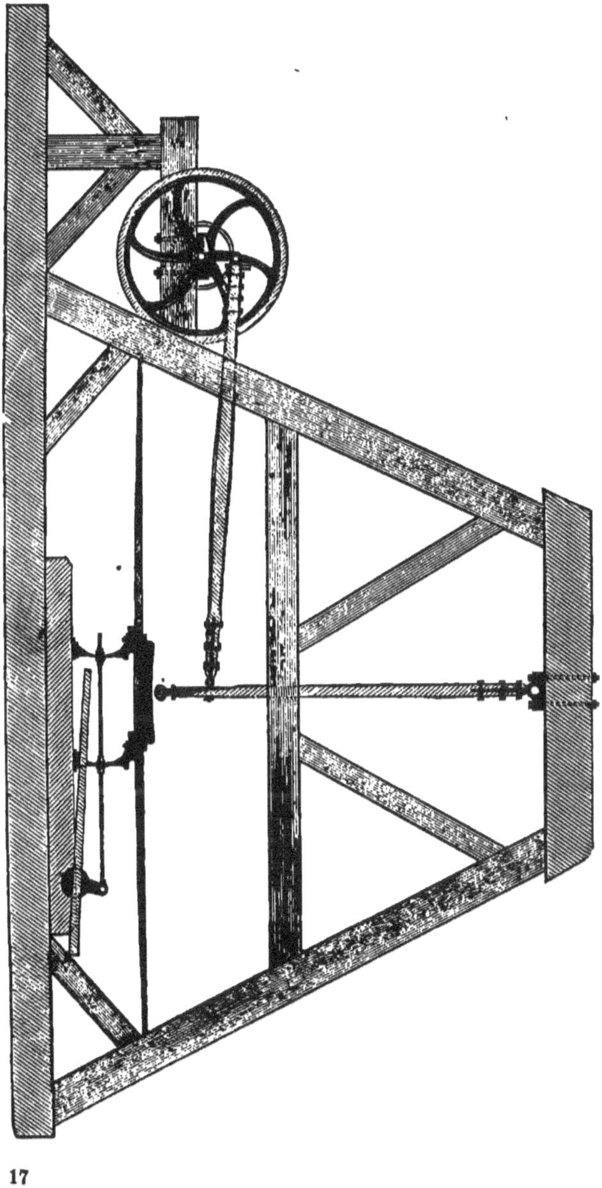

SOLE LEATHER ROLLER.

The representation on the opposite page shows the new form of dryer, now so generally in use. The building may be of any hight or shape ; the engraving is of a structure of four stories, besides the basement floor. This lower story should be reserved for steam pipes, or other means of heating. All the floors above are latticed, and when not obstructed by dampers, allow the free passage of air to the top opening. The draft may be regulated by "stops" or "dampers," either at the openings on the lower floor or at the top openings, or both, the object being to *absolutely control the currents of air, so that only so much, and at such periods as the attendant may desire, will air be allowed to pass.* If glass windows are used, as shown in the drawing, they are only to give light to enable the work to be done, but never to be opened for the purpose of admitting air or light to reach the leather, and when the work is done they are to be covered with a window shade, so that the leather may be excluded from the light during the whole drying process. This form of dryer is built on the plan of a chimney, and the laws of its action are precisely like those of that essential ventilator to our dwellings ; if currents are permitted to enter otherwise than at the bottom, counter currents will be formed in the dryer, and it will not "draw," but if the air is confined, then the difference between the temperature of the air at the ground and that at the elevation at the outlet will cause the current to rise, and rapidly just in proportion to its hight. Hence, to get a satisfactory self-acting turret, the building should be as high as possible. For more specific suggestions regarding the economic use and management of these dryers see Chapter XII.

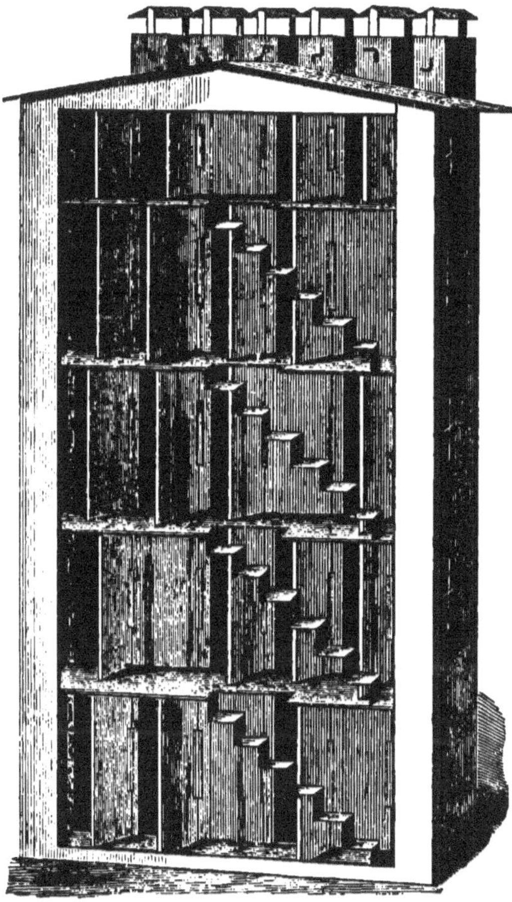

THE TURRET DRYER.

THE ALLEN & WARREN SPRINKLER LEACH.

This is a "percolating" process of leaching bark in contradistinction from the "press" system. Both methods are in very general use in this country. The Allen & Warren method is known as the "sprinkling" process, and is shown in the drawings. Many tanners, however, accomplish imperfectly a similar result by flooding their leaches with a limited quantity of liquor, allowing it to percolate through, and then flooding again and again, each time with about one-third of a vat of liquor. But if percolation is to be practiced at all, there can be no doubt that the Allen & Warren method is to be preferred.

This sprinkling process has been so much abused by tanners, and so much damage has been done thereby, both to the color and general quality of the leather, that a strong prejudice exists in the minds of many against this method of leaching; but it is believed that if the new bark is leached with cold or only warm water or liquor, then a concentrated and pure liquor can be obtained, while if the head liquor is heated up to, or near, the boiling point, and is then sprinkled on the new bark, an undue amount of coloring matter will be carried down and find its place on the leather, making both a dark color and a harsh texture and fiber. The practical working of these leaches proves that "strength" can be more concentrated with them than by any other means; hence, for extract manufacturing, they are serviceable, and we should judge that fully one-half of the hemlock tanners of the country are using them, without appreciating their good qualities or avoiding their bad ones.

The heater box, by which the exhaust steam from an engine will heat a constant stream of water or liquor without in the least reacting upon or retarding the power of the engine, may be seen in the two lower engravings. The sec-

THE SPRINKLER LEACH.

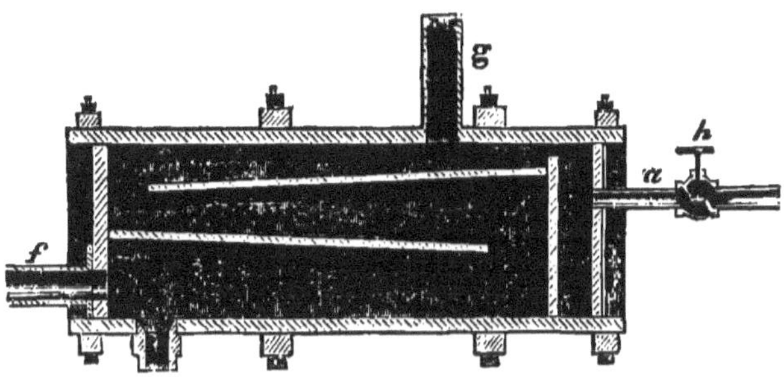

PAN FOR HEATING LIQUORS—SECTIONAL VIEW.

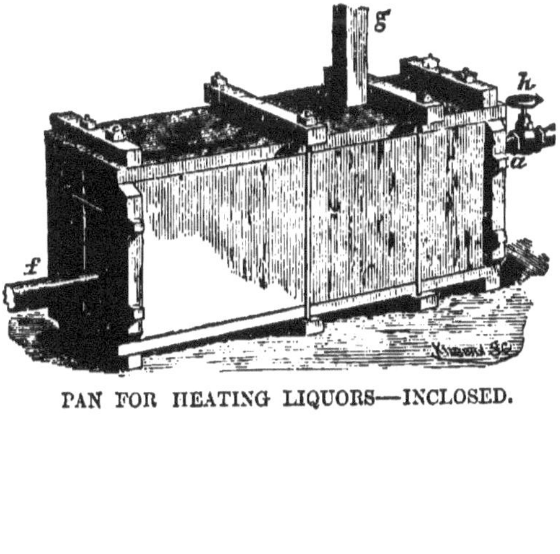

PAN FOR HEATING LIQUORS—INCLOSED.

tional figure represents two shelves, *c* and *d*. These are set on an incline, so that the water or liquor, which enters at *a*, fills the compartment *b*, until it overflows on the inclined shelf *c*, passes down on the inclined shelf *d*, and so finally out of the discharge opening *e*. The steam from the engine takes the opposite course, entering at *f*, passing through the liquor or water until it finds its way out at *g*. The steam thus passing will fully condense, and only moderately-heated air will pass off. The other drawing represents the enclosed box, with these shelves inside.

By this simple and inexpensive construction of plank a most effective heater is made, and by its use a constant stream of hot water or liquor can be run through on to the leaches.

If the most perfect heater possible is desired then small holes may be made in the shelves, in which case they should be made of copper plates—then the perforations can be small and effective; they will rain down small drops, condensing the steam most perfectly. This is one of the best and most effective improvements ever made in the heating of liquors. There is no patent claim on this improvement.

THE BARK CUTTING MACHINE.

The accompanying cut is designed to show the new machine for sawing tanners' bark. There can be no doubt that this machine does prepare bark for leaching better than any known bark mill of ordinary construction. It will saw damp or even wet bark with the same facility as if it were dry. If the bark is in large pieces it can be fed into the mill with facility, but if small then it will require a hopper or some other contrivance not yet perfected, certainly not practically introduced, to facilitate the feeding. It is claimed that a single mill will cut one cord and a half per hour. To do

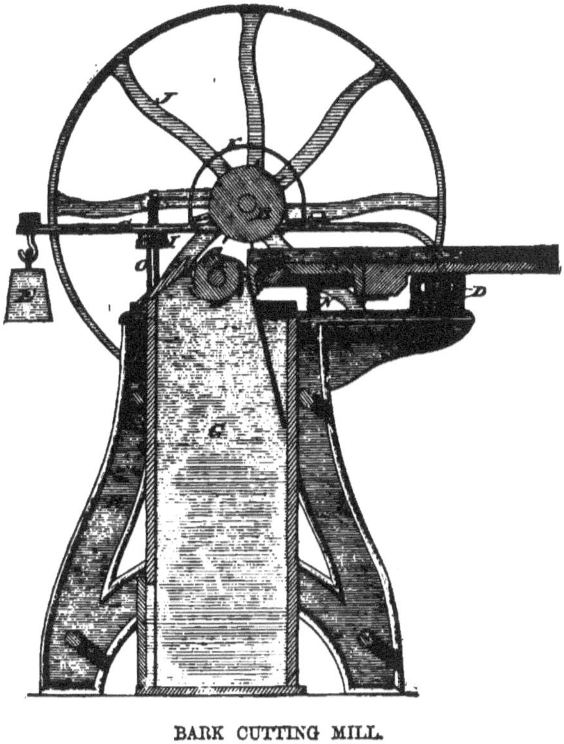

BARK CUTTING MILL.

this we should judge that the circumstances must be very favorable.

The high cost of this mill, $300, will probably restrict its use; it is, however, far better for tanners to pay even this price, unless it is possible to secure an equally good result by some more economical method. That other method is, in our judgment, by the use of a wire screen, as expressed elsewhere. There is no excuse for attempting to leach coarse and unevenly ground bark.

THE KEYSTONE BARK MILL.

The skeleton form of bark mill presented on the opposite page is probably the most skillfully manufactured machine of the kind known to the trade. We have taken special pains to describe its peculiar construction and its performance in chapter V, to which the reader is referred. Great pains are taken to make this mill run true; the ability to replace the surface segments, by either steel or cast iron surfaces, renders it one of the most serviceable, and, when we consider the safety coupling with which it is connected with the driving shaft, it is one of the most *durable* mills known. If tanners either cannot or will not adopt the wire screen, then we say this mill will grind more uniformly than any mill offering to the public, but it will not grind wet bark any more than will other cast iron mills.

THE HOWARD SCRUBBER.

As the drawing indicates, this scrubber consists of two skeleton drums, with projecting arms, into which are fastened birch brooms, closely held. The drums are rolled together by the cog gearing as seen on the side. When ready for work the box or covering marked A is dropped down, and through the slit or opening, G, the sides are fed. The feeding is per-

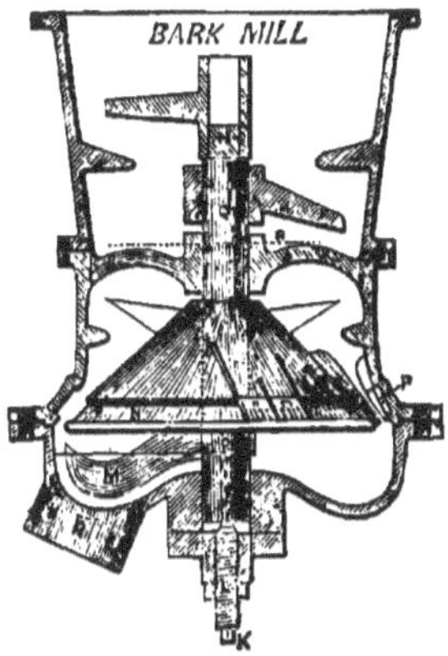

ALLENTOWN BARK MILL.

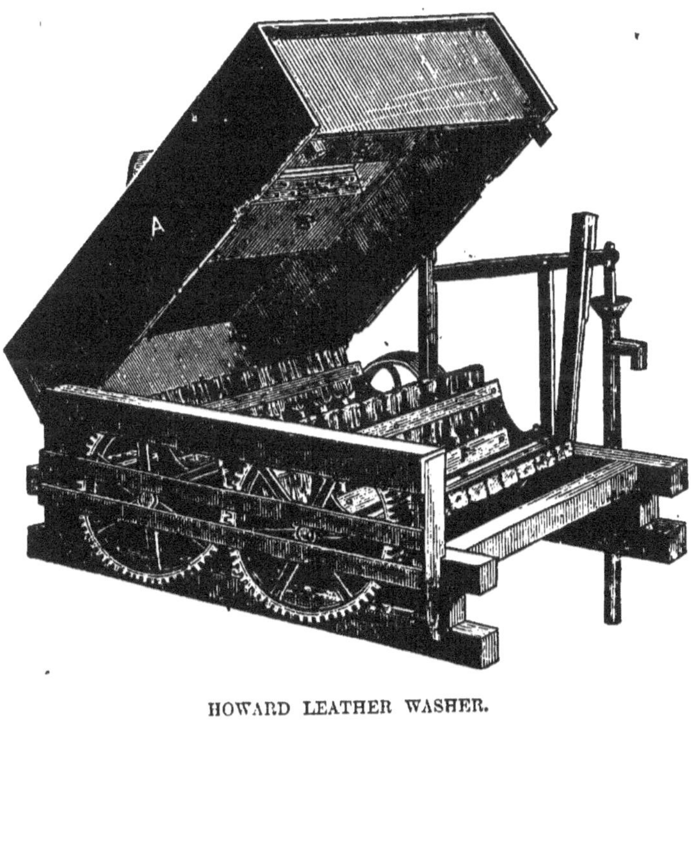

HOWARD LEATHER WASHER.

formed by allowing one end of the side to pass down between the revolving brushes; if the side is firmly held and allowed only to pass down slowly the surfaces will be thoroughly scrubbed, and well done just in proportion to the time allowed. To do the work well each side must be passed through twice, once from the head and once from the butt. One of these machines with two men to attend it will scrub about four hundred sides per day. It is simply justice to say that the "drum scrubber," elsewhere described, is taking the place of this Howard machine in some of our large hemlock tanneries, and tanners will do well to examine the merits of both before adopting either.

THE SALEM WET TAN PRESS.

This lever roller serves a most useful purpose for squeezing out the remaining ooze from spent bark, previous to burning it. Just to the extent that water or spent ooze is taken from the bark is the bark rendered serviceable to burn. It is now demonstrated that Thompson's patent for burning water, (which is equivalent to saying that wet tan will give more heat than dry tan), is a fallacy. But whether the power lost by squeezing out the water is not the equivalent of power lost in drying out the water in the oven is a question never yet determined by actual experiment.

This Salem machine is fully as efficient as the French and English machines for the same purpose, and seems more simple in construction. It is regarded as serviceable where the tanner requires more power than he can gain by the use of wet spent tan burned in an ordinary oven; but all sole leather tanners have an excess of tan, and, therefore, can never require this wet tan squeezer.

THE SALEM TAN PRESS.

THE LOCKWOOD AUTOMATIC LEATHER SCOURER AND HIDE WORKER.

On the opposite page is an illustration of this new invention, patented July 26th, 1876. The machine is novel in construction, simple in its movements, and ingenious to the extent of almost running itself, as the weight of a man's finger can guide the scourer over the surface of any kind of thick or thin leather. The patentee claims it to be a machine complete in itself, independent of building or framework; it can be set up without bolting or bracing, is durable, being made of metal of the utmost strength, and with air cushions which relieve the working parts of thrusts and strains. The machine can be set at any angle with the line of shafting, and belted on either end from above or below. Only from one to three horse power is required to run it, according to the thickness of leather being dressed. It occupies but little more space than an ordinary currier's table. It seems almost automatic in its movements, and is capable of the widest range of work, from the lightest to the heaviest; will scour, set out or glass, and can be made to take a slow or quick stroke, a long or a short one, making the most perfect stroke attainable with the smallest loss of motion, which is effected by the epi-cycle and cam combined.

THE LOCKWOOD LEATHER SCOURER.

THE FITZHENRY SCOURING MACHINE.

This machine has been so universally adopted, not only in this country but throughout Europe, that it is perhaps doubtful whether it should not be excluded as one of the *old inventions*, too well known to be entitled to a place in this volume, which claims only to present novelties, or such machines as are confined in their use to localities, and are comparatively new. But since within a few years we have another, and within a few months still one more, competitor for the honors of recognition, all three are given in contrast.

THE FITZHENRY LEATHER SCOURER.

THE BURDON SCOURER.

This practical and very serviceable machine is mostly useful in scouring out *bloom*, while it at the same time softens and cleanses the grain. It is useful for harness leather and fair leather curriers, but has not so far proved a success in scouring sole and belt leathers, which require more vigorous treatment. For calf, sheep and goat it should answer a most valuable purpose, and for grain leather it must prove a very efficient machine.

The high cost of the machine has no doubt prevented its adoption more generally by the trade. Had the improvement fallen into other hands, there can be no doubt that it would have come into general use before now.

THE BURDON LEATHER SCOURER.

THE STUFFING WHEEL.

The accompanying drawing presents one of the commonest forms of the stuffing wheel, now in very general use in this country. Through the opening door may be seen wooden pins on which the leather falls in the revolutions of the wheel, and they also serve the purpose of carrying upward the sides or skins, and generally agitating the fiber. This wheel may be driven by a belt, as shown in the drawing, or by cast iron gear.

The general methods of using this wheel have already been discussed in the preceding chapters. It was first practically introduced by Mr. Charles Korn, as early as 1856 to 1860. It had its origin in Germany, but there a barrel or hogshead served the purpose, while here special wheels of permanent construction were first introduced into the currying shop of Mr. M. M. Schultz, in the year 1860, at Sparrowbush, N. Y., under the direction of the original introducer.

This wheel has revolutionized the character of the upper leather of the country, which is now made soft and yielding, whereas before it was hard, the oil and tallow being now fulled into the center of the fiber, while under the old method, which applied the stuffing on the flesh side, it only penetrated a little beyond the surface, leaving the main body of the leather dry and unaffected. Leather prepared for this wheel should be in the same condition as to dryness as if stuffed on the surface in the ordinary way—neither too wet nor too dry.

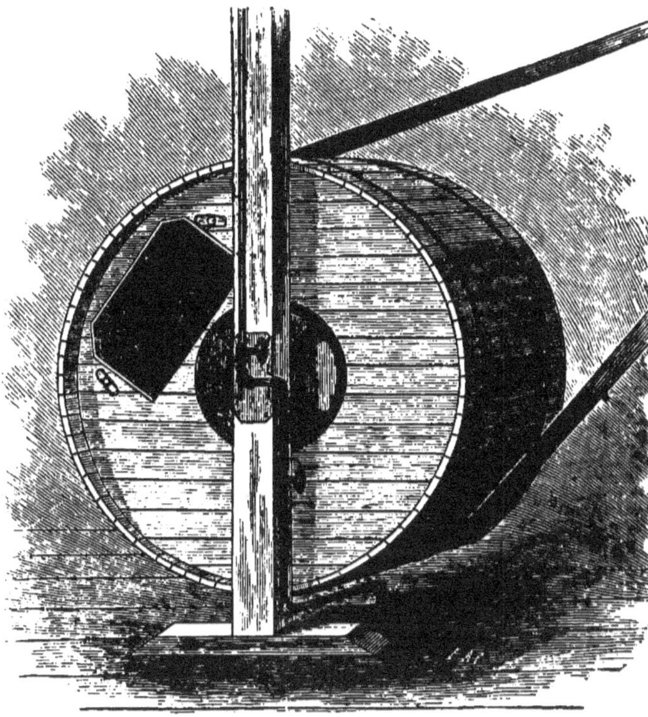

THE STUFFING WHEEL.

CHARLES KORN'S WHITENING MACHINE.

The accompanying illustration shows Mr. Korn's whitener, with the inventor operating it. The action of the machine is precisely that of a currier's knife in the hands of the workman by the ordinary beam process. The knives are fastened to an endless leather belt, and set diagonally, so that when the cut is made on the beam, as it passes down in front of the operator, there is a sliding and diagonal cut. The knives (for there may be as many as you please, certainly three or four) are cleared on their edge by one moving automatic finger, and by an automatic hand the edge is sharpened. This latter performance is the most complete success, performed by the most ingenious piece of mechanism, the writer has ever seen. The inventor claims that he can, with this machine, do the work of four men, and do it better than by hand work.

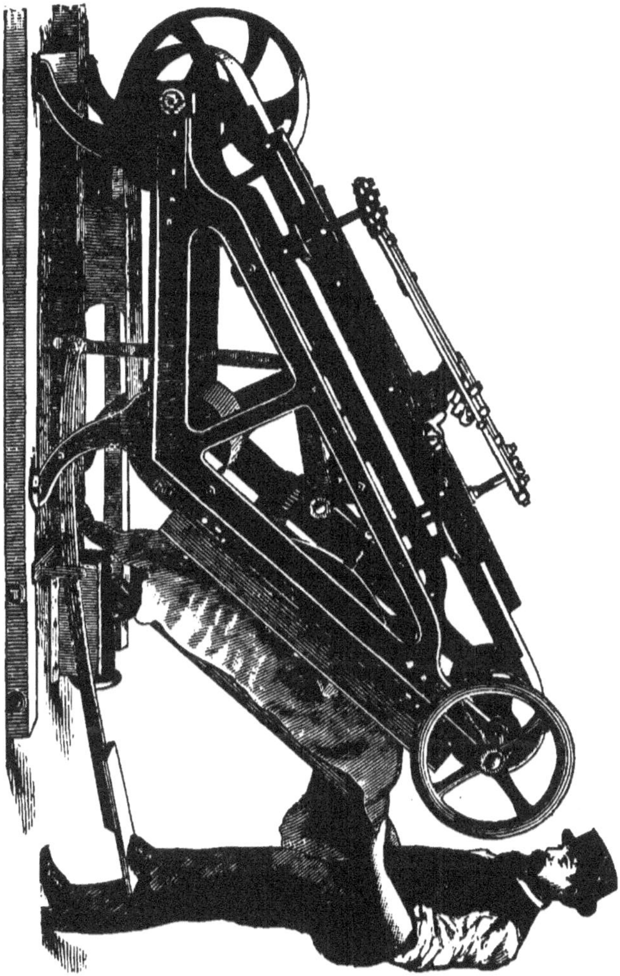

CHARLES KORN'S WHITENER.

THE "UNION LEATHER WHITENING, BUFFING AND SKIVING MACHINE."

Herewith is presented a representation of a recently invented machine for whitening, buffing and skiving leather. It is simple in construction, easily kept in order, requires from four to five horse power to run it, according to the thickness of the leather, and requires only an ordinary operator to perform as much work in one day by the use of the machine as is usually done by four whiteners. The quality of work is uniform and equal to that of the best skilled hand labor. The knives or blades make clean cuts, leaving no nap, and the leather, when finished, is smooth in the flanks, bellies and necks. The cylinder contains thirty-two knives or blades, inserted spirally, and a sharpener is attached, which travels forward and backward across the edges of the blades. The cylinder revolves 2,780 times per minute, and the pendulum swings to and from the operator at a speed of 90 per minute.

UNION WHITENER AND BUFFING MACHINE.

FISK'S WHITENING AND BUFFING MACHINE.

The inventor claims that this machine can and will do its work much more economically and even better than can be done by hand labor. Among its claimed advantages are the following: "It is small and compact," "runs with small amount of power," and "the table has a convex bed, and thus bends the surface of the leather from rather than toward the cutter"—"consequently," says the inventor, "the square edges of the blades take hold of the leather with very little pressure, and make a light, clean cut." Some certificates of those who have used the machine assert that wax leather, measuring from nineteen to twenty square feet per side, can be whitened by it in from one and a quarter to one and a half minutes, thus making about forty sides per hour. This, no doubt, is an extreme performance.

The general resemblance of this machine to "the union leather whitening machine," presented on another page, will strike all who compare their general structure. The difference consists in the working of the table, or the feeding of the knives. The knives themselves, and the form of sharpening them, may be considered identical. There can be no doubt that this machine does good work, and far more economically than it can be done by hand labor, so that it will probably share the patronage of all tanners and curriers who profess to study the economies of their trade.

FISK'S WHITENER AND BUFFING MACHINE.

"THE UNION LEATHER SPLITTER,"

shown on the opposite page, is too valuable to omit from any classification of American leather dressing machinery. It was the pioneer of all the improvements made. While we believe there are yet patents maintained on some unimportant parts of the machine, as a whole all patents have long since expired.

This machine has revolutionized the currying and finishing of leather in America, and is destined to do so in every other country. No currier can afford to carry on his business without a splitting machine of some kind, and this union machine is both the cheapest and simplest in construction.

THE LAMPERT HIDE WORKER.

This eccentric machine is well designed for unhairing hides or skins. The drum "A" revolves on the axis "B" at the center, rendering it unnecessary to move the hide sideways, but by a slight effort of the hand the drum is rolled from right to left, or left to right. The hide is drawn toward the workman as the parts become finished, whether the machine is used as a scourer or unhairer, for it is capable of doing any work which can be done with the worker. The stone or slicker is held by a steel spring, which in turn is propelled by the arm attached to the balance wheel. The machine should be seen in operation to be fully appreciated.

Eccentric motions are now used very extensively in several departments for the finishing of leather. Grain leather of all kinds is diced, and morocco and sheepskins are glazed and polished, as well as diced, by machines which use eccentric motion, with modifications of form and action.

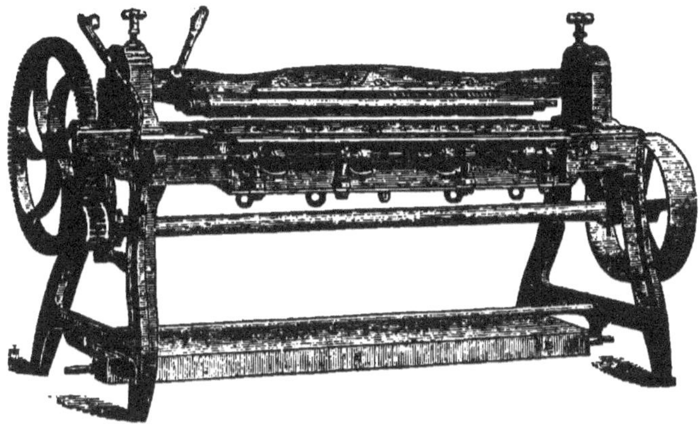

UNION LEATHER SPLITTER.

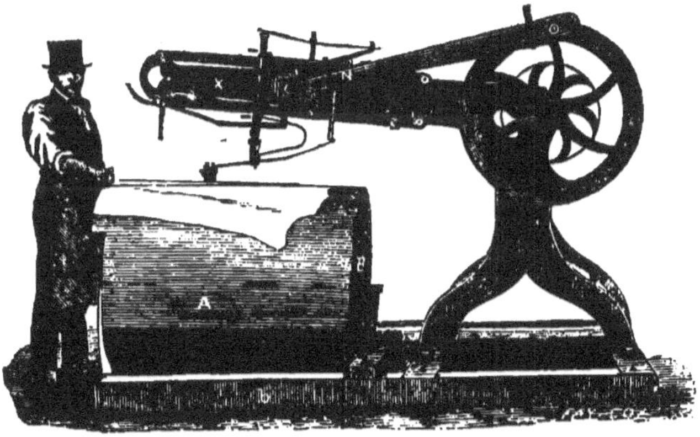

HENRY LAMPERT'S HIDE WORKER.

OUTLINE OF HIDE WITH TRIM USUALLY MADE.

The cutting of the hide in forms has become so common, from the demands of the belt, harness and shoemakers, that we have availed of the accompanying cut to locate each part by a designated term. This nomenclature conforms in part, if not altogether, to the English classification. "*A A*" are the bellies; "*B B*" are the bends; "*C*" and "*D*" are shoulders. Sometimes the shoulders are cut so as to include the neck, and in that case both "*C*" and "*D*" are cut in one piece, but it is more common to cut "*C*" as the shoulder. This trim will be varied by the fact as to whether the hide is a "cut" or "stuck" throat. The term "offal" applies to all the parts outside of the bends marked "*B B*," but the pieces are separately and specifically named as above.

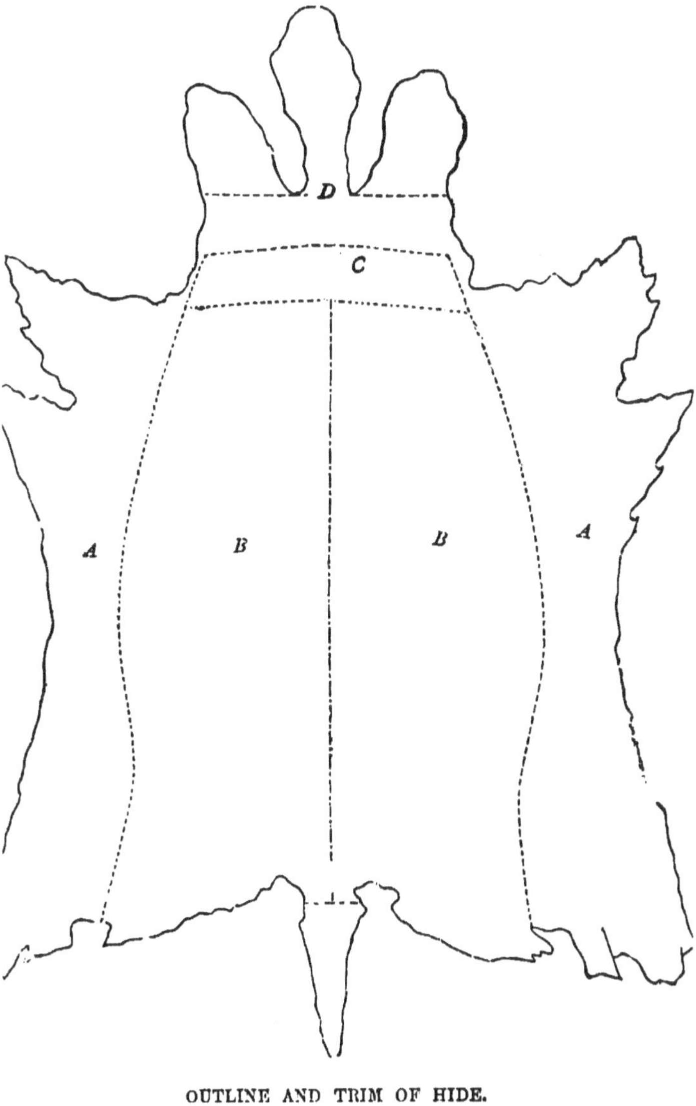

OUTLINE AND TRIM OF HIDE.

TANNERS' AND CURRIERS' TOOLS.

On the following page is a representation of a case showing the principal kinds of tanners' and curriers' tools in use. The center piece is a moon knife, and around this the others are arranged, in most attractive form. The two long, thin, straight knives, extending from the center toward the right and left hand upper corners, are the recently introduced German fleshers, which have of late attracted so much attention.

TANNERS' AND CURRIERS' TOOLS.

THE HOYT FURNACE AT WILCOX, PENN.

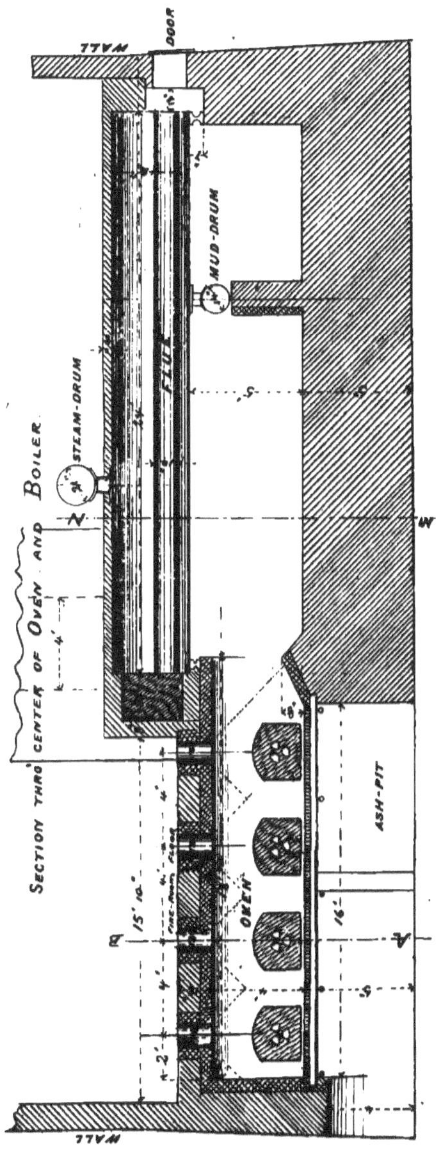

Plate II.

THE HOYT FURNACE AT WILCOX, PENN.

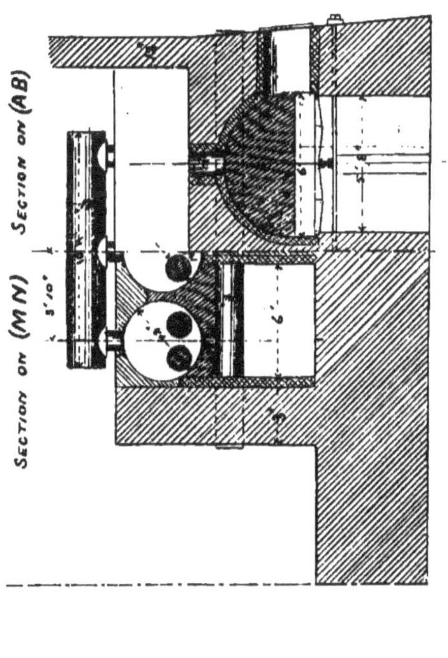

Fig. 1

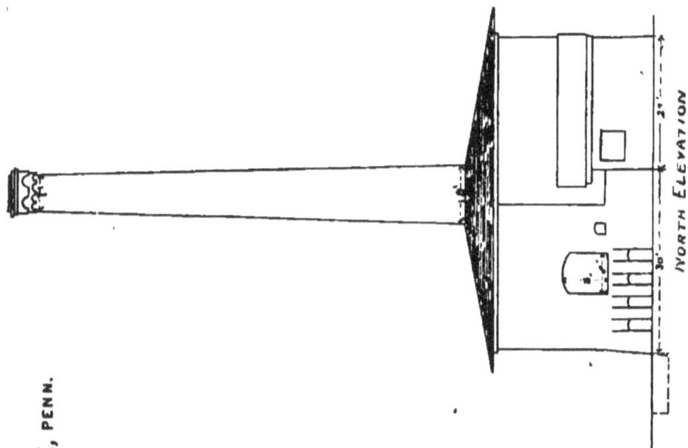

Fig. 2

THE HOYT FURNACE AT WILCOX, PENN.

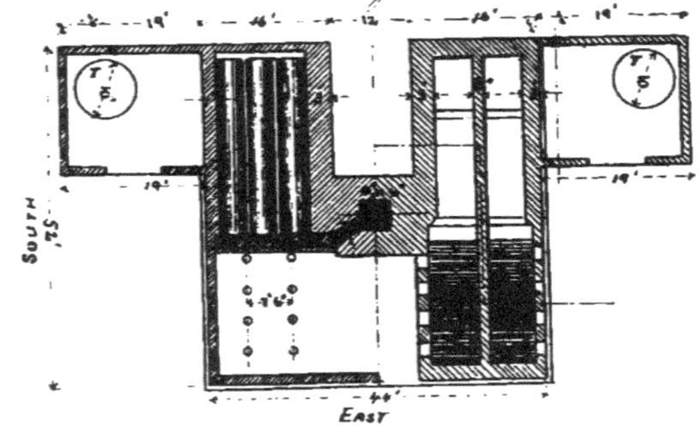

Fig. 1.

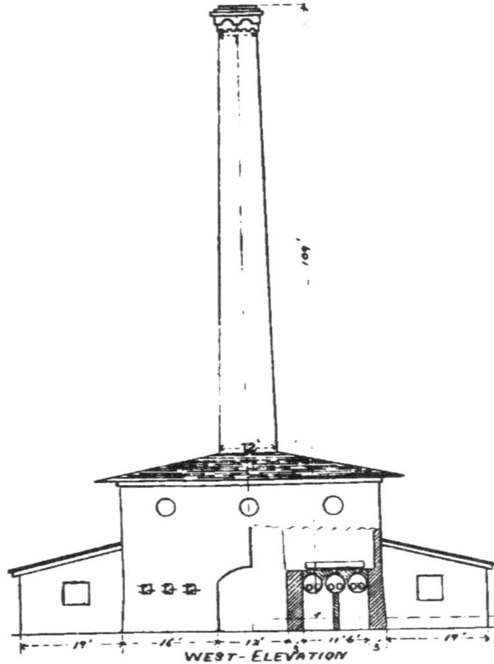

Fig. 2.

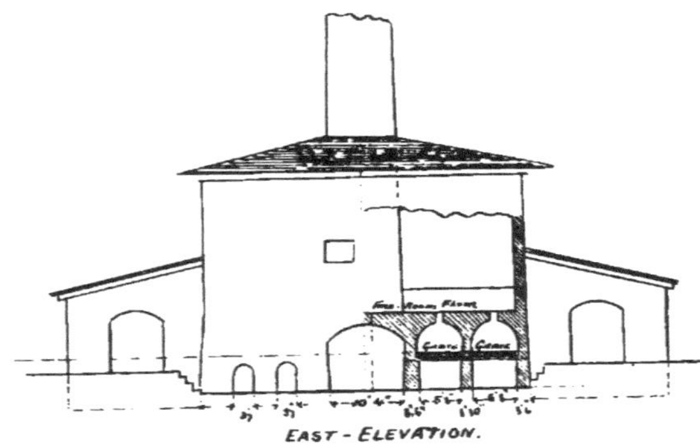

Fig 3.

Plate III.

Plate IV.

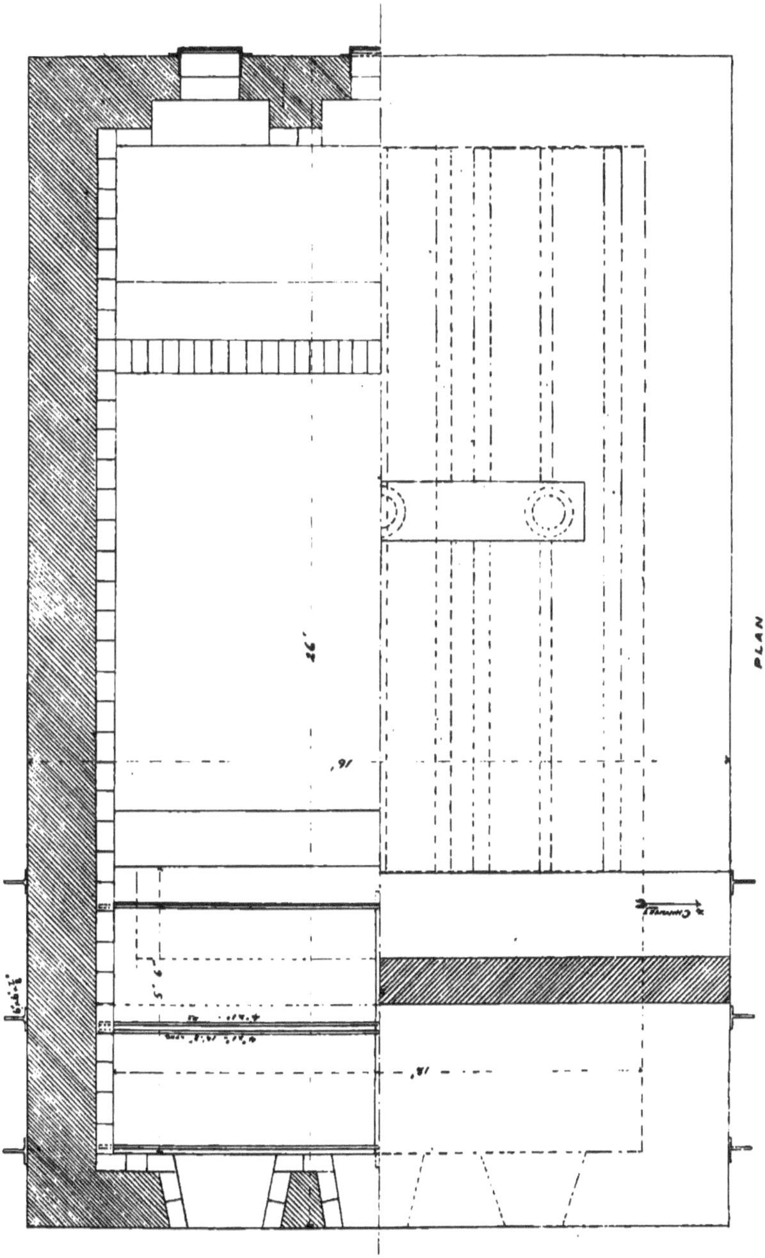

DESIGN FOR FURNACE FOR BURNING "WEST-SPENT-TAN"

DESIGN FOR FURNACE FOR BURNING "WEST-SPENT-TAN"

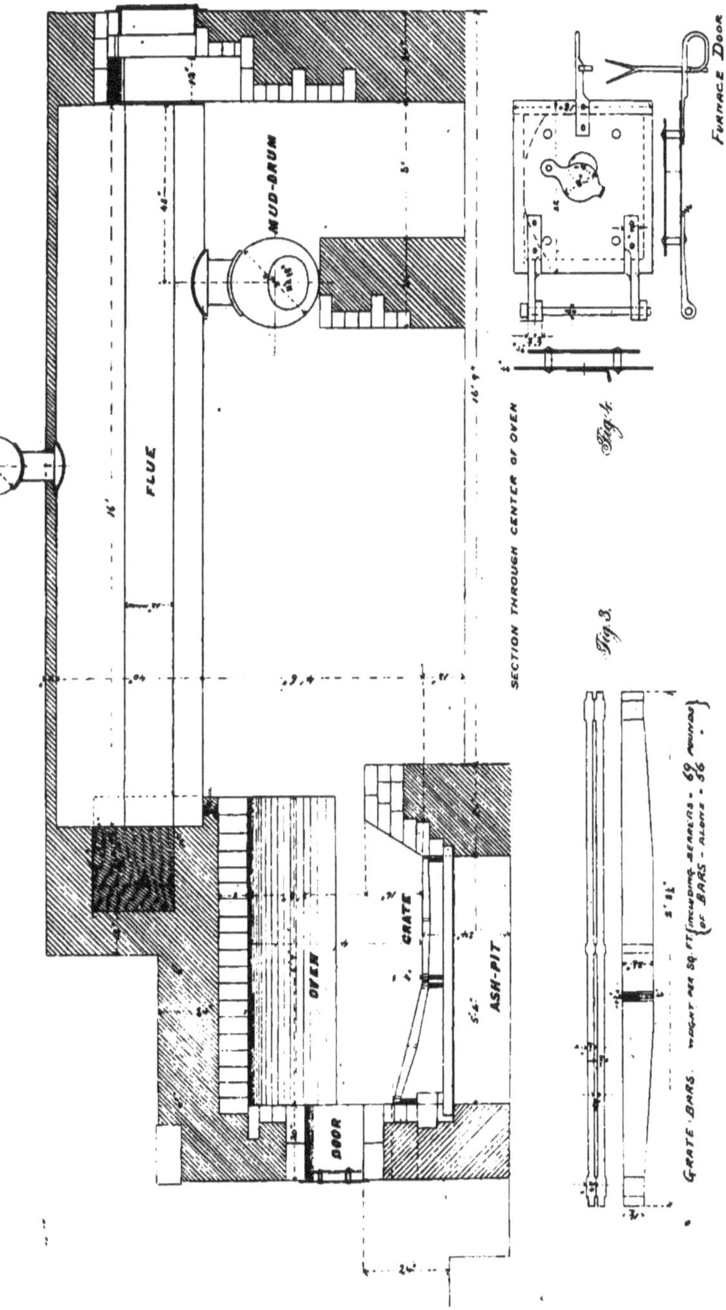

DESIGN FOR FURNACE FOR BURNING "WEST-SPENT-TAN"

Fig. 1.

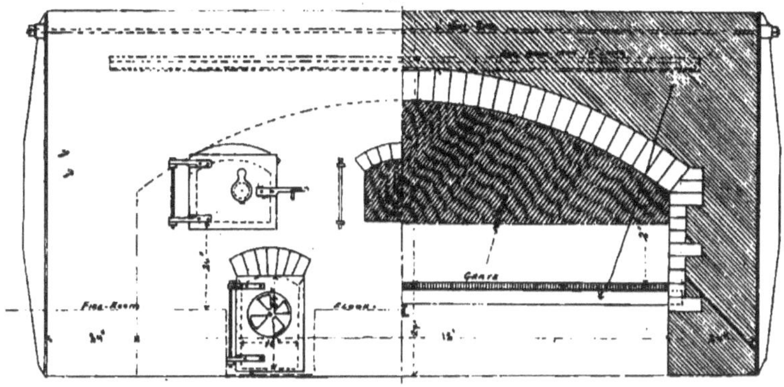

ELEVATION & SECTION THROUGH OVEN.

Fig. 2.

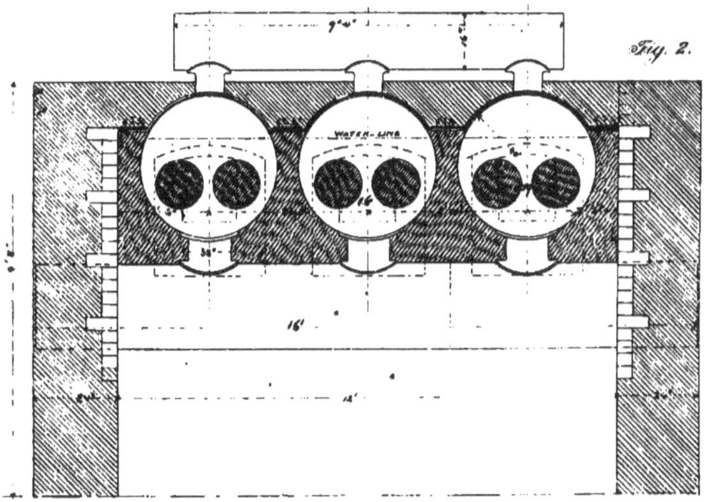

SECTION THROUGH BOILER.

ADAPTED TO BURN { 1/4 CORD AND TO FURNISH 1600 POUNDS OF STEAM
OR { 1/3 " " " " 2600 " " " } PER HOUR.

{ BEING THE SIZE RECOMME FOR A SOLE LEATHER TANNERY TANNING 20,000 SIDES
WHEN THE MACHINERY RUNS 12 HOURS A DAY OR FOR DOUBLE THAT CAPACITY (40.000
WHEN THE MACHINERY RUNS 24 HOURS A DAY

DIMENSIONS

GRATE-SURFACL	66	SQ. FEET
HEATING-SURFACE	670	" "
CROSS-SECTION OF FLUES		" "
DO DO CHIMNEY		" "
HEIGHT OF CHIMNEY	80	FEFF

INDEX.

A

Acid, use of in handling.............. 76
  in old liquors.................... 83
Agitation of the fiber of hides and skins promotes rapid tanning.183
Alkali for removing white spots..... 90
Ammonia in old limes................ 32
Army officers, prejudice against hemlock leather..................107

B

"Baggy" leather made by vacuum tanning.........................182
Bark dust........................... 51
  to suppress......................139
Bark elevators choking.............139
Bark, extractive and coloring matter with tannin in................ 55
Bark grinding...................... 46
  it should be ground fine and even 47
  grinding damp or wet bark...... 50
  screening ground bark.......... 50
  proper degree of fineness....... 51
  hemlock........................156
Bark rossing.......................141
  oak............................159
Bark mills—the "double grinder".. 48
  the Allentown mill..........49, 265
  the saw cutting mill.........51, 262
  crushing machine................ 52
  capacity of different mills....... 53
  speed to run mills.............. 53
Bating with warm water and hen manure........................ 34
  after liming.................... 81
Beam work.......................... 38
Blacking, flesh....................221
"Black rot" in sole leather......... 89
Bleaching leather...................100
Birch bark tanning.................169
Boot and shoe industry of the United States according to the census of 1870.......................244
Box vats...........................122
"Buffalo" vats.....................121
Buffing and whitening machines 41, 276, 278, 280
Buff leather.......................210
Burning wet spent tan, the first success in........................147
Burning wet tan................47, 224

C

Calfskins, French and German......194
the best goods made sent to Great Britain and the U. S........195
care in assorting the green stock.195
soaking and milling.............196
breaking the nerve.............197
liming and unhairing............198
bating.........................199
coloring and handling...........200
laying away, grain to grain......200
excluding the grain from the air.201
flesh shaving....................202
scouring........................203
stuffing........................204
slicker whitening...............205
boarding........................205
kept in russet state.............206
palm oil used in currying.......207
Census of 1870, boot and shoe industry of the United States........244
leather industry of the United States in 1870...............245
Chimney, dimensions of for wet tan ovens........................229
Clay for a tannery foundation........119
Clewer, James', method of leaching.. 57
Climate and food, effect of on hides..189
Color, a natural hemlock desirable...105
Color of hemlock leather improved by sumac.......................101
Coloring and resinous matter in bark 56
Coloring of leather to be done in the handlers.....................108
Concentrated liquors obtained by the sprinkler leach............... 66
Conductors of liquor decay rapidly when above ground..........124
Cost of boiler for 20,000-side tannery.......................231
Cost of furnace at Wilcox, Pa.......240
Cost of tanning, the............169, 243
"Cropping" leather 41
"Cuir" color......................106
Currying and finishing.............218
Cutch.............................163

D

Dampening before rolling........... 97
Degras in making stuffing...........221
Divi divi.........................165
Domestic hides of the Eastern, Middle and Western States............192
Draining after scrubbing............ 96
Drum wheel handler................ 70
Dry hides, table showing cost of tanning.........................243
Drying in the turret dryer.......... 97
saving in time and labor and improved color made..........111
size of building necessary.......113
how steam should be admitted..113
openings for light and air.......113
elevator for raising the leather..115
influence of light and heat......115
conveying leather from the yard to the drying lofts......136

E

East India kips.....................191
Elevator for raising leather to the drying loft...................115
"England" wheel, the.... 73
Experiment with water and tan liquor 63
in tanning New York City hides.. 85
in tanning by hydrostatic pressure.........................177

F

Ferguson, Thomas T., experiment at Sparrowbush, N. Y., on vacuum tanning.......................179
Fertilizing liquids..................153
Flesh blacking.....................221
Fleshing, cost of thorough work..... 38
lime slaughter stock............ 39
before liming................... 40
sweat stock..................... 40
the German flesher.............. 41
Fisk's whitener and buffing machine.280
Furnaces, directions for constructing for burning wet tan...........224

G

Gambier, or terra japonica..........162
its use as a substitute for bark...164
Gallic acid, sole leather raised by ... 80
German flesher, the................ 41
German sole leather................ 80
Glue stock, saving and caring for the.149
Grain and buff leather..............210
difference between that made from tanned leather and that made from green hide.......211
stamping.......................212
strength of....................212
suggestions for manufacture of..213
Grain, nature of the...............103
great care necessary to make it perfect.......................104
English custom of striking out the 108
Grate surface necessary in wet tan ovens.........................227
Grease on hides to be removed by an alkali.................. ...30, 91
how to purify and cleanse.......219
Green hides, table showing cost of tanning.......................243

H

Hair, cattle and calves', saving the..150
cleansing and preparing for market.........................151
Handling........................... 68
the hand reel................69, 252
the rocker handler...........69, 254
the drum wheel handler......... 70
a method of raising packs....... 71
tub wheel handler.............. 72
the "England" wheel........... 73
mechanical power.............. 74

Handling—The "Cox" rollers....... 74
on frames....................... 75
use of acid in.................... 76
for upper and harness leather... 83
Hand reel, the..................... 69
Hanging hides in sweats............ 27
Heating liquors, pan for............260
Heating surface necessary in wet tan ovens.........................228
Hemlock bark, a fair colored leather from 56,101
white and red.....................156
weight of a cord of different kinds of.........................156
growth of in different sections of country.......................158
Hemlock leather made to imitate oak leather..........................107
Hides, selection and classification.... 17
assorting at each stage of the tanning process.................. 17
softening in soak and mill....... 18
breaking the nerve.............. 19
kinds to be worked at the same time......................... 21
sweating......................... 23
hanging in sweat pits........... 27
liming............................ 31
bating............................ 33
trimming before tanning........ 43
skinning the cheeks and throats. 44
handling and plumping......... 76
laying away..................... 84
experiment with N. Y. City...... 85
cells of filled with water and not air..............................178
species and growth of...........188
healthy and well grown.........188
differences in at various seasons.189
climate and food, effect of on....189
Hide mill, the.......................250
Hide worker, Henry Lampert's......282
Hook marks........................ 91
Howard leather washer, the.....95, 264
How much leather will a cord of bark make.............................160
Hydrostatic pressure, experiment in tanning by.......................177

I

Improved breeds of cattle make hides thin and spready..............190
Instantaneous combination of tannin with gelatine................. 92
Insurance, old and new rates for tanneries...........................138

K

Knoederer's vacuum tanning process.181
Korn, Charles', whitener...........276

L

Laying away, time required for sole leather........................ 84
gives better color than handling. 85
Laying away as practised in Europe.. 86
in Great Britain................ 87
weight of liquors on last layer... 88
grain or flesh up................ 91
Leaches, round or square...........126
construction of..................128
capacity of to be proportioned to tannery.........................122
brick and cement for in England.133
Leaching........................... 55
proper degree of heat to be used. 57
the double leach................ 58
the press leach.................. 58
how it is worked............ 59
loam packing for.........60, 119
construction of............. 61
operation of................ 62
number of times bark is washed with............ 64
the Allen & Warren leach....65, 260
concentrated liquors obtained by. 66
the McKenzie leach............. 66
revolving leach.................. 67
Leather interest of the United States according to the census report of 1870..........................245
Leather should be dry when stuffed..222
Leather splitter, the Union..........282
Liming after sweating............. 28
when vitriol is used............. 29
to unhair........................ 31
kind of lime to use.............. 31
preparation of the limes......... 31
old limes........................ 32
time required for................ 32
Prof. Lufkin's process........... 34
the "Buffalo" method............ 35
handling in limes................ 36
Liquors, feeding hides with strong.. 82
Loam foundation desirable for tanneries..........................119
Loam packing for leaches........... 60
Location of tanneries...............118
placing the buildings to reduce fire risks....................137
Loss of tannin in making leather....173

M

Milling after sweating............... 28
Morocco vs. "pebble grain" leather..212
Myrabolams 166

N

Nerve, breaking of the, in hides and skins..................19, 40, 197

O

Oak bark............................159
the rock oak, white oak, and red oak.........................160
growth of oak trees.............161
Oiling before drying................ 97
Old liquors, acid in................. 83
One story buildings best for tanneries.........................134

P

Palm oil used by European curriers..207
Patents for tanning inventions......176
Paste.................................221
Press leach, the...................... 58
"Puddling in" the vats.............123

Q

Quick tanning process.............176

R

Reel for handling leather........69, 252
Refuse of tanneries—Utilization of..146
Rocker handler, the............69, 255
Rollers for handlers................ 74
Rolling, proper condition of the leather for.......................... 98
effect of the first................ 99
a second rolling necessary....... 99
the American leather roller.....256
Rossing bark.......................141
difficulty of separating the ross..142
will it pay to ross bark for export?........................144
"Rounding"........................ 43
Russia Leather, original color of.....106

S

Safety coupling for a bark mill...... 49
Scouring—the Lockwood scourer....268
the Fitzhenry scourer...........270
the Burdon scourer.............272
Screen for ground bark............. 50
Setting the vats and leaches.........120
Sheepskins, effect of vitriol on....... 79
Skins—Breaking of the nerve in French and German calfskins. 20
Selection before tanning........ 20
Liming 34
Fine texture of French, German and Swiss calfskins..........193
Size paste...........................221
Soap blacking.......................221
Soda ash, with lime, for unhairing... 35
Soda ash or sal soda for removing white spots................... 91
Softening hides..................... 18
Soil and climate, effect of on bark...156
Sole leather, German............... 80
Splitting machine, the............. 210
Sprinkler leach, the................ 65
Steam power to be conveyed by steam in pipes rather than by shafting..........................137
Streams, "manufacturing" and "culinary".......................118
Strength of liquors for handling..... 83
for laying away................. 84
for last layer................... 88
Stuffing wheel, the................. 27
Sugar of lead and sulphuric acid for bleaching leather............100
Sulphur, with lime and soda ash, for unhairing 35
Sumac baths for making a fair color.101
Sweating, in Europe............... 23
by steam....................... 24
construction of cold sweat pits... 24
to control the temperature of the pits........................ 25
grease and salt on hides, stopping the action of the sweating process................. 29
improved sweat pits............248
"Sweet fern".......................166

T

Tan burning...................47, 224
the first practical success in.....147
Tanneries, construction of.
Leaches... 126
Framework and location of buildings.........................134
The turret dryer................111
Plans, foundations, etc..........117
Tannin, difficulty of separating from extractive matter............. 55
instantaneous combination of with gelatine.............. 92
Tanning materials..................153
Tanning, the cost of................169
in Europe.....................174
Tan press, the Salem.................266
Tawing hides and skins..............186
Temperature of sweat pits.......... 26
of water in wheeling limed hides. 33
of limes....................... 35
of water for leaching with the press leach................. 57
of the stuffing wheel............219
Time of tanning by Knoederer's vacuum process.................183
Tools for tanners and curriers......286
Transferring packs, a method of..... 71
"Trim" of leather adopted by English tanners................... 42
Trimming of hides...............43, 284
Tub wheel handler, the............ 72
Turret dryer, the..............110, 258

U

Union whitener and buffing machine, the.........................278
Union leather splitter, the...........282
Upper leather tanned by a cord of bark.........................170

V

Vacuum tanning....................177
leather made "baggy" by.......182
Valonia.............................165
Vats to rest on the log conductors..121
"Buffalo,"......................121
box.............................122
Vegetable oils for currying..........207
Vitriol, the use of lime on hides raised by.......................... 28
its use in Great Britain.......... 29
its employment in handling..... 76

Vitriol—Difficulty in determining how much to use............ 78
should not be used on sweat stock........................ 79

W

Warm water to be used in soaking hard hides.................... 18
Washing after coming from the layaways........................ 94
Water in cells of hides..............178
Wheel scrubbing.................... 95
Wheel stuffing......................219
Whitener, Charles Korn's............276
the Union......................278
Fisk's.........................280
White spots in sole leather.......... 89
Wilcox, Pa., cost of wet tan furnace at..........................240
Working hides and skins to break the nerve......................... 40

APPENDIX.

REPORT

OF

THERON SKEEL, C. E.,

ON THE

COMPARATIVE ECONOMIES

OF

BURNING WET SPENT TAN

BY

THE VARIOUS DETACHED FURNACES NOW IN USE

BY THE TANNERS OF PENNSYLVANIA

AND NEW YORK.

NEW YORK:
OFFICE OF THE SHOE AND LEATHER REPORTER.
17 SPRUCE STREET.
1876.

BURNING WET TAN IN DETACHED FURNACES.

In order to explain and justify the great particularity with which this series of experiments have been conducted, it is proper to say that in the suit of Black vs. Thorne, in which was involved the merits of the so-called Thompson patent for burning wet tan bark, saw dust, &c., both Profs. Silliman and Thurston insisted upon the theoretical and practical superiority of the Thompson furnace over those in public use at the time of this patent. Prof. Thurston was compelled to admit, however, that methods outside of the Thompson patent were quite successful, and gave practically all the power required, but thought, from his observation and experiments at two tanneries, that the Thompson furnace would give about thirty per cent. better results from a given amount of wet spent tan than the furnaces outside of the patent.

As this result was known by all practical tanners not to be true, they naturally felt anxious to have the experiments made with more care than Prof. Thurston had pretended to exercise. He guessed and estimated too much, and weighed and measured too little, to carry the conviction of his theories, as against the known results in actual practice. Hence it was that J. B. Hoyt and J. S. Schultz, representing the tanners, sought for some expert that should carefully retry the experiments so imperfectly performed by Prof. Thurston, and they were greatly aided in this endeavor by Mr. B. F. Isherwood, Chief Engineer of the U. S. Navy. The letter received from him so fully sets forth the grounds of his commendation that it is here inserted with the correspondence and report which follows.

New York, August 17, 1875, 111 East 36th Street.

Dear Sir:—I take the liberty to write you in regard to some experiments which I understand you contemplate, and which I hope you will have made, on the evaporative efficiency of wet tan bark burned in the furnace of a steam boiler. The subject is one of great interest in engineering, which is my apology for thus troubling you; and the problem should be solved in so complete and appropriate a manner as to remove forever the doubts now resting on it.

In my opinion there is no person of my acquaintance so thoroughly qualified to conduct such experiments, and ascertain their results, as Mr. Theron Skeel. He has all the necessary qualifications of an education, both mathematical and practical, and he is an expert experimentalist. He is thoroughly versed in the subject, and his report will command a respect and confidence not easy to obtain for that of any other. Whatever experiments Mr. Skeel makes will be devised in a manner to defy hypercriticism as to the propriety of their results, and his literary ability will enable him to present them so clearly, forcibly and full that they will be understood and accepted by all.

Should you conclude to have these experiments made, I volunteer the advice that you allow Mr. Skeel to have full power in devising and conducting them. By so doing I am confident you will obtain such results, and in such a manner, as will forever settle the questions at issue. The worst extravagance will be to so limit the experiments as to cause the least shadow of doubt to remain. I hope, in the interest of industrial science, that you will furnish Mr. Skeel the means to do justice to both himself and the subject.

Your obedient servant,
B. F. ISHERWOOD,
Chief Engineer U. S. Navy.

Jackson S. Schultz.

New York, Aug. 20, 1875.

Mr. Theron Skeel:

My Dear Sir:—Within a few days I have received a letter from B. F. Isherwood, Esq., Chief Engineer of the U. S. Navy, in which he recommends you as a suitable person to test the evaporating power of wet tan bark when burned in furnaces in the front of steam boilers.

In view of your conceded ability, in common with other tanners who are interested in the subject above referred to, and particularly as developed by the controversy arising out of the suit now pending on the Thompson patent for burning wet fuels, we desire that you at once proceed to test, by a series of experiments at several tanneries, the comparative merits of the various methods, but particularly those of Thompson, Hoyt, Crockett, and other such modifications of these as in your judgment may tend to throw light upon the general subject embraced in the "*consumption of wet fuels in detached furnaces or ovens.*"

* * * * * * * * * * * * *

Our wish and instructions to you are, that you make your tests in the most thorough manner, and, when you have finished your work, that you report to us in writing.

Very truly,
(Signed) J. S. SCHULTZ, on behalf of J. B. Hoyt and others.

REPORT.

86 Irving Place, New York, Dec. 30, 1875.

Jackson S. Schultz, *Esq., Nos.* 63 *and* 65 *Cliff Street, New York*,
J. B. Hoyt, *Esq., and others :*

Gentlemen—I submit the following report of the experiments made in accordance with your letter of instructions preceding.

The experiments were made on three forms of furnaces designed to burn spent tan in the condition in which it comes from the leaches. The object of the experiments was to determine in which kind of furnace the hot gas developed by the burning of a given quantity of wet tan would evaporate the most water in the boilers attached. The furnace which excels in this respect may be said to be the most *economical* of *bark*.

This must be distinguished from commercial economy, which depends upon the value in money of the spent tan, the cost of the furnace and of attendance and repairs. In the present state of the art about one-half of the spent tan is thrown away in all sole leather tanneries, for they only use about one-half of all the tan made to make the steam necessary for the whole tannery. Under these circumstances the moiety of the tan has no value in any case, and in some locations is worse than valueless, for the tanners are prohibited by law from dumping the tan into the stream, or where there is no stream to dump it into (as is sometimes the case), they are obliged to either cart it away or to erect an additional furnace especially to burn it. The term economy as used in this paper refers to *economy in the use of bark only.*

The economy of any furnace (the term furnace including the

whole combination of ovens and boilers) evidently depends upon:

1. The total heat which would be developed by the perfect combustion of the quantity of tan fed into the oven.

2. The portion of the heat that is developed by the combustion that takes place in the furnace.

3. The portion of the heat developed which is utilized by the boilers.

The first condition depends entirely upon the bark, and is independent of the oven or boiler.* The second condition depends almost entirely upon the oven and is independent of the boiler. The third condition depends both upon the furnace and boiler.

In order that any set of experiments should be conclusive in themselves without reference to any conditions but the quantity of tan burned and of water evaporated, it is necessary that a set of ovens should be built of each kind and connected to exactly similar boilers, and that these ovens shall be fed at the same time with the same quantity of tan from the same leach, and that the pressure of steam and the temperature of the feed water shall be the same in each case. Such experiments would be comparative, but are not practicable. In the absence of exactly similar conditions, it is necessary to consider (so far as we are able) all the variations.

The experiments described in this report were made in 1875 as follows:

Sept. 22—Crockett furnace, Stevens' tannery, Great Bend, N. Y.
Sept. 24—Crockett furnace, Wells' "Southport" tannery, Webb's Mills, N. Y.
Sept. 26—Thompson furnace, H. F. Inderlied, Brackneyville, N. Y.
Sept. 29—Thompson furnace, Weed's tannery, Binghamton, N. Y.
Oct. 6—Crockett furnace, Wells' "Southport" tannery, Webb's Mills, N. Y.
Oct. 16-17—Hoyt's furnace, Wilcox Tanning Co., Wilcox, Pa.
Dec. 9—Crockett furnace, Wells' "Southport" tannery, Webb's Mills, N. Y.
Dec. 10—Crockett furnace, Wells' "Southport" tannery, Webb's Mills, N. Y.

The plans and dimensions of the ovens and boilers and the detailed description of the experiments are given in the appendices to this report.

FUEL.

The fuel was in all cases spent bark of hemlock trees, called "tan." The trees are felled while living and stripped of their bark during the summer season. The bark is piled up in the

*If the tan were so thoroughly leached and dried as to be fed into the furnace nearly in the condition of *perfectly dry wood* it is probable that the combustion would be more nearly perfect, with the same supply of air, than if the same furnace were fed with imperfectly leached bark.

woods after being stripped, and during the winter hauled into the tannery and piled up again until needed for use.

The bark is peeled from all sizes of hemlock trees down to about 6 inches in diameter, the largest being about 30 and the average perhaps 15 in diameter. Good average hemlock land will peel about 12 cords of bark to the acre. One man will fell, peel and pile about 1½ cords of bark per day.

The bark when originally peeled from the trees, at the same season, has probably nearly the same composition in all localities. This has been found to be true of wood, and therefore inferentially of bark. In that condition, then, a given weight of bark would probably correspond to a given quantity of heat in all cases. If, however, it is not burned at once, it dries more or less according to the length of time and the circumstances under which it is stored. The time of storing varies in practice from a few weeks to several years, and while sometimes it is stored in large piles and exposed to the weather, at others it is stored in small piles under sheds. After being stored a certain length of time it loses all the water except about 14 to 16 per cent., leaving the balance containing 84 to 86 parts of dry wood and 14 to 16 parts of water.* If more thoroughly dried than this by exposure in a hot kiln, it will on exposure to the atmosphere reabsoıb an amount of water varying from 14 to 16 per cent. with the hygrometric condition of the air. Dry wood, in the sense used in this paper, is understood to be wood that has parted with all the water it will when exposed to dry air at a temperature of 110° Centigrade or 230° Fahrenheit.

The unit of measure of the bark in this condition (before being ground,) called "chip bark," is among tanners the cord, being a pile of bark 8 feet long by 4 feet wide by 4 feet high. The weight of this volume of bark is variously estimated at from 1,800 to 2,240 pounds, and probably varies within these limits with the manner of packing and the weight of water contained.

Before leaching the bark is ground in a machine called a "bark mill." The bark in its original condition consists of an outer shell of hard dry substance, analogous to "cork," and an inner

*I am informed by a gentleman who has had considerable experience in drying wood for manufacture into spools that green wood if cut into faggots and exposed to the air in sheds protected from the weather, would lose in one year all the water it contained except about 15 per cent., as compared with kiln dried wood at 150 deg. F. The kiln dried wood would reabsorb this 15 per cent. when exposed to the air.

layer of a substance containing more sap and more like woody fiber. In the first process of grinding the outer shell grinds much finer than the inner. In some tanneries the whole product from the mill is sifted in a revolving screen, and the coarser part, consisting mainly of the large pieces from the inner layer, is sent back to the mill and reground.

The weight of a cord of ground unleached tan (being determined by weighing the contents of several boxes containing exactly 25 cubic feet each) the bark being shoveled into the box, not packed, and struck off with a straight edge, was found to be at the various tanneries as follows:

Wells'	2,582 pounds
Brackneyville	2,353 pounds
Wilcox (fine)	2,418 pounds
Wilcox (coarse)	2,284 pounds
Average	2,351 pounds

The water contained in the last two specimens as compared with the same dried at 110° C. was found to be:

Wilcox (fine)	18.1 per cent
Wilcox (coarse)	17.0 per cent

And therefore the weight of a cord of dry ground unleached bark:

Wilcox (fine)	1,980 pounds
Wilcox (coarse)	1,895 pounds
Average	1,938 pounds

But at Wilcox all coarse ground tan is sent back to the mill and reground, and therefore the weight of a cord of fresh ground bark will be the mean of the coarse and the fine, or 2,418 pounds, the percentage of water contained 17½, and the weight of the cord of dry bark 1,995 pounds.

Probably no sensible error will arise if we assume for all cases the weight of a

Cord of dry unleached ground bark = 2,000 pounds.

In this condition probably the same weight of bark will develop the same amount of heat in all localities, and within the limits of practice the same is true of equal volumes.

After being ground the bark is "leached," that is, placed in a wooden leach having a perforated bottom and saturated with warm water for several days. This water percolates through the bark and carries off with it, if the process is continued long enough, *all soluble matter*.

After the soluble matter is supposed to be all out, the water is allowed to drain off and the residue, called "wet spent tan," is shoveled out of the leaches and carried to the furnace to be burned.

A cord of the bark in this condition has been in former experiments taken as a unit of measure of the amount of heat equivalent to the tan fed into the furnace. In order to test the accuracy of this assumption the whole quantity of tan burned in each experiment was measured and weighed, as in case of dry bark, by shoveling it into a box containing about $\frac{1}{8}$ cord, striking off with a straight edge, and weighing the contents.

The weight of a cord of *wet leached tan* under these circumstances was found to be at various tanneries as follows:

Stevens'	4,442 pounds
Wells' (1)	4,294 pounds
Wells' (3)	4,275 pounds
Weed's	4 270 pounds
Wells' (4)	4,260 pounds
Wells' (2)	4,225 pounds
Brackneyville	4,112 pounds
Wilcox	4,076 pounds
Average	4,244 pounds

The per centage of water lost by drying* at 110° C. was found to be at the various tanneries:

Wells'	63.4
Wells' (2)	62.8
Wells' (4)	62.3
Wells' (3)	61.5
Stevens'	61.5
Wilcox	61.2
Brackneyville	59.0
Weed's	55.1

*The average composition of bark dried at 80 deg. C. is given by M. Violet;

Carbon	48.6
Hydrogen	6.3
Oxygen	41.8
Ash	3.3
	100

If this bark had been dried at 110 deg. C., as was done in the experiments, its composition would have been changed by the loss of 5 per cent. of water, and would have been:

Carbon	51.2
Hydrogen	7.0
Oxygen	39.3
Ash	3.5
	100

Or of 51.2 per cent. carbon, 1.2 per cent. hydrogen, 44.2 per cent. of hydrogen and oxygen in the proportion to form water, and 3.5 per cent. ash.

The weight of the water which one pound of the bark in this condition would evaporate from 212 deg. F., computed by allowing 15 pounds per pound of carbon and 64 pounds per pound of hydrogen, will be:

Carbon	51.2@15 =	7.68
Hydrogen	1.2@64 =	0.77
		8.45

Of this 55-100 of pound of water are latent in gas, leaving available heat 7 9-10

And consequently the weight of the dry portion of the cord :

Weed's	1,917
Wilcox	1,582
Wells' (4)	1,606
Wells' (3)	1,646
Brackneyville	1,686
Stevens'	1,710
Wells' (1)	1,572
Wells' (2)	1,572
Average	1,661
Average weight, rejecting Weed's	1,625

In order to try the effect of packing on the weight of the cord, several boxes were weighed containing tan packed by a man (weighing at Wells' about 150 pounds, and at Brackneyville 180 pounds) stamping it down as fast as shoveled in, the top being struck off with a straight edge as before. The weight of the cord was as follows :

Wells'	5,327
Brackneyville	5,490
Average	5,405

The per cent. of water contained being as before, the weight of the dry portion will be :

Wells'	1,950
Brackneyville	2,251
Average	2,100

At Wilcox the whole contents of two leaches were weighed. The weight of a cord at the density of packing in the leach was found to be :

Wilcox	5,219

And the weight of the dry portion :

Wilcox	2,025

In the experiment marked Wells' (3) and Wells' (4), the tan was used during the first part of the day from the bottom of one leach and during the second part from the top of another. The following exhibits the weight of a cord taken from various parts

pounds from 212 deg. Probably no sensible error will arise if the thermal equivalent of one pound of bark dried at 110 deg. C. is taken at 8½ pounds of water evaporated from 212 deg. F.

The thermal equivalent of a cord should be in water evaporated from 212 deg :

$$2,000\times8.5=17,000 \text{ pounds.}$$

Prof. Johnson has made some experiments on dry pine wood giving 7¾ pounds of water from 212 deg. F. No estimate of the water contained is given, but assuming it to have been 20 per cent. (and it could not have been less), there results from his experiment :

Pounds of water evaporated from 212 deg. F. by pound of pine wood (dry).........9.38

In his experiment, 1 36-100 pound of water were latent in gas, leaving available heat, 8.02.

of the leach, the whole capacity being 10 cords, the bottom cord being called the 10th:

1st day,	8th cord in leach,	48 hours draining,	weight	4,262 pounds
1st day,	10th cord in leach,	56 hours draining,	weight	4,306 pounds
2d day,	1st cord in leach,	24 hours draining,	weight	4,262 pounds
2d day,	2d cord in leach,	32 hours draining,	weight	4,211 pounds
2d day,	3d cord in leach,	38 hours draining,	weight	4 257 pounds
2d day,	4th cord in leach,	44 hours draining,	weight	4,290 pounds

A sample of bark to determine water was not kept separately for each cord. The weight of a wet cord seems steadily to increase from the top toward the bottom, but not very rapidly, the whole increase being less than 2 per cent. The masses of bark were broken up by shoveling out of the leach, and were all measured at the same density, these figures having no reference to the tightness of packing in various parts of the leach.

The extreme variations of weight of the dry portions, from 1,917 at Weed's to 1,572 at Wells', would at first seem to point to an error in measurement. The following exhibits the difference:

	Wells'.	Weed's.
Weight of cord of wet tan	4,294	4,270
Weight of dry portion	1,572	1,917—difference 345, Weed's most
Water	2,722	2,353—difference 369, Welles' most

*The dry green bark, having the same composition in both cases, is, according to M. Violet, made up of:

1. Of a material called cellulose, which has always the same composition.

2. Of an incrusting material which is richer in carbon and hydrogen, and of which the principal constituents are resin, gum, starch, sugar, glucose and tannin.

3. Of mineral matter or ash.

The soluble constituents are wholly or partially extracted, according as the process of leaching is more or less complete. It is possible, then, that the soluble portion was nearly all extracted from the Wells' bark and but partially extracted from the Weed's bark, the pores being filled with water in both cases, but there being more empty pores to contain water in Welles' bark.

If this should be true, the balance of the soluble matter may be extracted by thorough leaching, and the dry portion remaining should have been nearly the same weight in every case.

In order to test this theory a given weight of wet bark was taken, the sample from each tannery boiled for 2½ hours, the

*Annales de Chimie and de Physique, vols. XXIII. and XXXII.

liquor filtered, and the balance dried at 110° C. as before. The result of this experiment was as follows:

	Original weight of dry cord from leach.	Loss of weight by boiling 2½ hours.	Final weight of leached dry cord.
Stevens	1,710 pounds	84 pounds	1,626 pounds
Brackneyville	1,649 pounds	113 pounds	1,536 pounds
Weed's	1,917 pounds	408 pounds	1,509 pounds
Wells' (1)	1,572 pounds	93 pounds	1,479 pounds
Wilcox	1,582 pounds	98 pounds	1.484 pounds
Wells' (2)	1,572 pounds	148 pounds	1,424 pounds
Average	1,667	157	1,510

The variations of the weight of the dry boiled portion may be due in part to the fact that the soluble matter is not yet all out, for they were only all boiled the same time, and not until they ceased to lose, as would have been the final test. Enough, however, was done to show that so far as the extremes are concerned, Weed and Wilcox, the measurements are probably correct.

A specimen of the unleached bark from Wilcox's was soaked in cold water for 36 hours and boiled for 6 hours, the liquor filtered, and the residue dried at 110° C. as before:

	Original weight of cord dry unleached bark.	Loss of weight by leaching and boiling.	Final weight of cord boiled dry bark.
Wilcox (fine)	1,980	341	1,639
Wilcox (coarse)	1,895	366	1,529
Average	1,937	354	1,589
Screened and reground bark at Wilcox	1,995	365	1,630

This bark would probably have lost more if the process of leaching had been continued for six days, as in practice, in place of 42 hours.

Probably no sensible error will arise if the weight of a cord of *thoroughly* leached bark from the tannery be taken at 1,400 pounds, when dried at 110° C.

It appears, then, from these experiments that the average weights are as follows;

	Cord of average green ground bark.		Cord of well leached wet bark from tannery.	
Soluble matter	500 pounds	20 per cent	000 pounds	00. per cent
Insoluble matter	1,500 pounds	60 per cent	1,500 pounds	35.7 per cent
Water	500 pounds	20 per cent	2,700 pounds	64.3 per cent
	2,500 pounds	100 per cent	4,200 pounds	100 per cent

The experiments do not show what the 500 pounds of extractive matter consists of, but only that it is material soluble in hot and cold water.

The proportion of water does not increase with the size of the lumps, as was found by drying a specimen of very large lumps from cord. The water contained was shown to be exactly the same as in the fine.

A cord of tan thoroughly leached and saturated with water weighs more and contains more water than a cord imperfectly leached.*

A cord of fine ground tan weighs more than a cord of coarse ground, as a barrel of meal weighs more than a barrel of corn.

THE CROCKETT FURNACE.

The Crockett furnaces experimented upon consist of an oven of fire brick, constructed near the boiler, and having a set of cone grate bars of cast iron. The tan is fed in through doors in the front.

The depth of the furnace is limited, as in the ordinary furnace for coal, by the distance to which the fireman can easily throw the tan, to 6½ feet, but may be as wide as is necessary to give the grate surface, and varies in practice from 6 to 12 feet in width.

The fire is started (as in every case) with dry wood, and continually replenished until the brick work is hot, after which no more wood is necessary. The frequency of feeding depends upon the rapidity of consumption. When burning at the rate of 7 pounds of dry tan per square foot of grate per hour, each portion of the furnace requires to be fed with fresh tan every three-quarters of an hour at least. If there are several doors, each door has to be opened every three-quarters of an hour. At Welles' furnace there were three doors, and some door was opened every 15 minutes. The time of feeding under these circumstances was about 1½ minutes, so that the doors were open one-tenth of the time. If the fuel burned regularly over the grate, the feeding could be less frequent.

The crown of the furnace is a segmental or elliptical arch, and the distance from the crown to the grate varies from 1½ to 2½ feet in different ovens. The fire bridge is at the back of the grate, and is a continuation of the back vertical wall of the ash

*The specific gravity of the material extracted from the bark is nearly 1½ times that of water, so that the 500 pounds of material extracted would be replaced by 330 pounds of water. The weight would be as follows:

	Before leaching	After leaching
Soluble matter	500	....
Solid matter	1,500	1,500
Water mechanically held between particles	2,250	2,250
Water in pores originally filled with soluble matter		330
Weight of cord of wet tan	4,250	4,080

These are nearly the weights of cord at Wilcox and Weed's.

pit. It rises about 1½ feet above the grate, and extends entirely across the oven. The arch over the oven is generally from 2 to 3 feet longer than the grate, and forms the support for the front end of the boilers.

The boilers are generally horizontal flue boilers, with two flues; the products of combustion after leaving the furnace, passing forward under the boilers, returning through the flues, and finally passing in front to the chimney through a sheet iron connection. The chimney is generally of sheet iron, and is sometimes set directly over the boilers, and sometimes carried on a brick foundation.

The main difficulty in keeping a uniform fire is the comparatively small weight of fuel in the furnace at one time. If the fire is carried on an average 9 inches thick, there will only be in the furnace at one time (if the grate is 6½ feet long by 12 feet wide) 800 pounds of combustible. At the rapid rate these furnaces are generally fired, being, in the case of Wells' (1), 800 pounds dry tan per hour, this would only last *one hour*. A furnace burning coal at the rate of 10 pounds per square foot per hour would contain coal enough to last 6 hours. The conditions of a tan bark fire, then, in the Crockett furnace, fed every hour, may be imagined by supposing a coal fire fed every six hours.

STEVENS' FURNACE AT GREAT BEND.

The performance of this furnace during the experiment indicates the economy of the Crockett furnace when forced to the utmost. During the experiment the number of pounds of dry tan burned per hour was:

Per square foot of grate surface	10.70
Per square foot of heating surface	1.61
Per square foot of the cross area of flues	258 00

In order to consume this tan the damper and ash pit doors were kept wide open and the furnace fired every ten minutes.

The method of firing was first to fill the holes that had burned through, and then to cover the whole surface with a layer of tan about 3 inches thick.

The tan thrown in first forms a uniform layer over the surface. It appears to undergo a species of distillation, during which the water is evaporated, the tan maintaining its original brown color.

Just before firing, the furnace was a dull red over about half its surface, the color fading out into a black toward the front, and the surface of the tan was burned through into holes wherever the original tan was thinnest or the fire hottest, but mostly where the grate had broken away and left large crevasses through which the tan fell into the ash pit. At the instant before firing, the temperature of the gases leaving the furnace was scarcely sufficient to melt silver, and there was no smoke coming from the chimney, but occasionally a few sparks.

The fireman would now throw in as described the charge of about 300 pounds of wet tan through both doors alternately.

Immediately on closing the doors the destructive distillation of the tan commenced, under the combined influence of the heat radiated from the surface of the arch above and of the hot embers on the grate beneath. By looking in through some of the crevasses in the front of the furnace hundreds of little jets of white smoke could be seen issuing from the whole surface of the tan.

During the process of distillation, which lasted about 3 minutes, the smoke and steam which was unable to find an outlet into the chimney issued in small quantities through the crevasses in the brick work into the fire room, or was forced back through the grate into tbe ash pit. A careful examination of the edges of the crevasses in the brick work showed them to be covered with a deposit of an oily substance, undoubtedly condensed from the smoke, showing that at this time the combustion was imperfect, and that the smoke issuing from them had contained combustible matter. During the process of most rapid distillation, no air entering the ash pit, *combustion must have been entirely suspended*, and whatever gases were generated must have passed away unconsumed. Toward the end the crown of the furnace becomes black all over, owing no doubt to the heat absorbed by the water in the tan.

Now there are signs of combustion commencing. These are, first, the cessation of the smoke from the crevasses into the fire-room, and little jets of yellow flame replacing the smoke in the furnace. They begin gradually, and in about two minutes cover the whole surface of the tan, which gradually grows red hot and consumes. From the time of the first appearance of the flame, the crown of the furnace gradually grows hotter, and just before firing is red hot over half its surface as before.

The appearance of the smoke during the first part of the cycle indicates the presence of combustible gas at that time. Unfortunately, the loss of heat by combustible gas does not cease when the flame commences, for although at that time some of the gas is consumed, probably a large part of it, forced from the tan by the heat of the embers below, and issuing into a comparatively cool oven, does not acquire sufficient temperature to consume, even though mixed with sufficient air. I consider that in the Stevens furnace, fired in this way, probably the whole effect of the gaseous portion of the fuel (about 25 per cent.) was lost.

This loss was aggravated by the very thin layer in which the bark was spread over the whole surface, allowing all the water to be evaporated in a few minutes, the volume of gas completely filling all egress from the boilers, thus arresting combustion and allowing the furnace to cool down below the temperature necessary for the ignition of the gases expelled during the subsequent process.

This loss would have been probably almost entirely prevented by closing the ash pit down so as to burn about half as much tan, or by putting in two more boilers, so as to have twice as much area of flue. In either case the effect would be to give an excess of area to the outlet over the inlet of the oven, so that the combustion need not cease while the distillation is taking place, in which case the temperature of the oven would be maintained nearly constant, and the gas consumed as fast as expelled.

It is also necessary for the combustion of the gas that there should be supplied an amount of air that does not pass through the incandescent tan on the grate. Unless this air is supplied the gas will pass off unconsumed.

This air is supplied practically by the holes burned in the fire. The tan, after being dried in the furnace, commences to burn away most rapidly where tan is thinnest or the draft strongest. The layers of bark around this hole do not slide down and fill it, but stand perfectly perpendicular, or even overhang a little. These holes generally form on the crest of the grate bars, and in a few minutes enlarge rapidly, and finally lay bare the whole crest. They act a very efficient and beneficial part if not allowed to increase too much, but check the fire and cool the oven if left too long without being filled.

A description of the performance of a Crockett furnace, when these holes were in one experiment allowed to form, and in another when they were kept from forming by filling up all hollows with fresh tan before the surface had burned away (marked Wells' 3 and Wells' 2), in which it appeared that with the same rate of combustion, and with the same temperature in the oven, the performance of the furnace in the case where the holes were kept open exceeded the performance when the holes were kept shut by *more than 25 per cent.*

In the case of the Stevens furnace at Great Bend, the main

cause of the failure in economy was undoubtedly owing to the *high rate of combustion at which it was forced.*

CROCKETT FURNACE AT WELLS' TANNERY.

This furnace was similar to the Stevens furnace, except that there were 3 boilers, each 42 inches in diameter and 22 feet 6 inches long, with two horizontal flues, 13 inches internal diameter. The grate was 6 feet 6 inches deep by 12 feet wide.

The boilers had been in use for 22 years and were said to be covered with scales on the inside and the brick work was full of fissures. There were many leaks around the sheet iron work connecting the boilers with the stack and in the back connection.

The oven was new and in good order. It had been recently built to replace a Thompson oven torn down, being connected with the same boilers. The grates were "cone grates," and were new and in good order.

In the first experiment, marked "Wells' (1)," the furnace was tested in the cendition it was found, the rate of combustion being a little more than enough to supply the steam necessary to run the tannery.

In the second experiment, marked "Wells' (2)," the air leaks in stone work and sheet iron work were closed, as far as practicable, with mortar, and the rate of combustion was reduced to the same as at "Weed's," being nearly 64-100 pounds of dry tan per square foot of heating surface per hour. All the holes in the fire were filled, as soon as they commenced to develop, with a shovel full of wet tan.

In the third experiment the same pains were taken to stop air leaks, same rate of combustion was maintained, only holes were allowed to form in fire, and the top of the oven was covered with wet tan, which was afterward burned as in the furnaces fed from the top.

In the fourth experiment same conditions were maintained as in last, except that cross section of flues were reduced to $2\frac{1}{2}$ square feet.

EXPERIMENT MARKED "WELLS' (1)."

The performance of the furnace during this experiment indicates the effect of forcing a Crockett furnace beyond its capacity, the rate of combustion being in pounds of dry tan per hour:

Per square foot of heating surface	0.98
Per square foot of grate surface	10.50
Per square foot of cross section of flues	147.00

Being more than 50 per cent. in excess of the rate necessary to allow perfect combustion.

The performance was injured by the cold air which leaked into the flues through fissures in brick work, and particularly by the cold air which had leaked into the chimney by the opening in the sheet iron work around front connection. An opening was found here after experiment was concluded exposing 36 square inches. The temperature of the gas leaving the flues, taken in chimney beyond this opening, was 580°, or 270° hotter than the steam, while at Wilcox, at nearly the same rate of combustion, where there was no leak, the temperature of the gas 700°, or 380° hotter than the steam. The inference would be that the temperature of the gas at Wells' (1) would have been 680° if no cold air had leaked into flue between the boiler and the thermometer. The proportion of air necessary to reduce the temperature 100° would be 16⅔ per cent. The weight of air found in the chimney was 21 6-10 pounds per pound of dry tan by chemical analysis, and, therefore, the air that passed through the flues was 18 pounds per pound of dry tan.

The air passing through the fire was still less than this by an amount that leaked into the flues en route from the furnace to the chimney.

The furnace was fed about every 10 minutes through *one door* at a time, thus making the weight of a charge about 300 pounds of wet tan, the balance of the hourly consumption being thrown in, in occasional shovel's full, to fill up holes.

No care was taken ts fill up holes burned in fire. These were so large, just before feeding, as to expose about one square foot of opening for passage of cold air from ash pit through holes in grate bars. The temperature of the gases leaving the furnace, as indicated by a piece of metal on the bridge wall, was just sufficient to melt silver at their hottest. This occurred about midway in the interval of time between the feeding and the burning down of the fire ready to be fed again. The surface of the arch over the furnace showed at this time a bright red over about half its surface, fading out toward the front. As the holes developed in the fire, the color died away, and was a *dull red* just before feeding. After feeding the color faded away and was invisible for a few minutes, gradually recovering to a bright red just before feeding again. The extremes of temperature of the furnace were then probably 1,000° and 1,900°, and the average about 1,500°, corresponding to about 15 pounds air per pound dry tan.

The general appearance of the furnace was as at Stevens', only less exaggerated. The period during which the tan was not burning was shorter, and the smoke was never returned into the fire room.

The large supply of air was probably mostly due to the holes allowed to burn in the fire. Probably a considerable portion of the combustible gas was wasted in this experiment. The heat unaccounted for in the gas, water and steam was 0.40 pounds of water evaporated from 212°, or about 5 per cent.

EXPERIMENT MARKED WELLS' (2).

After the completion of the experiment marked Wells' (1), and noticing the injurious effects of the holes burned in the fire, it was determined to try the effect of keeping the holes entirely closed, and at the same time to reduce the rate of combustion to that at Weed's.

The outside of the boiler was carefully covered with mortar to prevent any air leaking into the flues. The only air that did leak in was probably between the boilers and the chimney and had no effect on the economy.

The surface of the fire was carefully watched, and as soon as a hole began to develop it was closed by opening the furnace door and throwing as quickly as possible a shovel full of tan directly into it.

The furnace was divided into three sections, corresponding to the three doors, and these sections were fired every hour, the weight of the charge being 400 pounds of wet tan, the balance of the hourly consumption being used to fill up holes. The fire was kept about one foot thick. The temperature of the furnace was higher than during the previous experiment, and nearly the same as in the succeeding experiment, indicating a smaller supply of air. As no air could reach the chamber in which the gases were generated without first passing through the 12 inches of tan, there could but very little free air reach them at all, and they probably passed away entirely unconsumed, entailing a loss upon the furnace of about 25 per cent. The steam formed was only three-fourths of that in the succeeding experiment, or indicating a loss of 20 per cent.

EXPERIMENT MARKED WELLS' (3).

The results of this experiment probably are the maximum for this furnace in the condition which it now is.

The rate of combustion was the same as in last experiment, and the air leaks were carefully stopped, except those in the sheet iron work between the boiler and the chimney. The furnace was fired about every 15 minutes, the weight of the charge being about 300 pounds of wet tan, the balance of the hourly consumption being scattered in as the fire seemed to need it.

The temperature of the furnace was more nearly uniform, being at the highest sufficient to easily melt silver, and at the lowest a dull red, or about 1,200° and 1,900°, and the mean temperature of not less than 1,650°. I estimated the mean temperature from the comparative time the highest and lowest lasted.

The average opening of the furnace doors during the hour was about 7 minutes. The area exposed by the open furnace door for the inflow of air was about 300 square inches, while the average opening of the ash pit doors was about 100 square inches, or about one-third. It appears then that the furnace doors would have passed into the furnace 3 times as much air as the ash pit doors while they were open, but as the furnace doors would only be open about 7 minutes during the hour, would only admit about 30 per cent. of the air admitted through the ash pit. The temperature of the fire when the doors were closed being as high as in any furnace tested, shows that at that time the tan was burning with as small a supply of air as in any case, or 10 pounds per pound dry tan.

The weight of air per pound of dry tan, as determined by velocity of smoke, was 16 pounds. A considerable portion of this must have leaked in through the opening in the sheet iron work.

The temperature of the furnace is computed from heat :

Given up by products of combustion to steam, taking weight of air per pound of dry tan at 16 pounds, and adding temperature of chimney, 490 degrees.... 1,310°
From the thermal equivalent of tan, allowing 0.25 pound for radiation, and assuming products of combustion to take all heat, the supply of air being 16 pounds per pound dry tan.. 1,300°

But the mean temperature of furnace was found to be 1,650°, indicating that 25 per cent. of the air in the flues had not passed through the furnace, leaving the supply of air 12 pounds per pound of dry tan.

In this experiment the draft was controlled by nearly closing the ash pit doors. No smoke came from the chimney. The top of oven was covered with tan.

This was the most successful experiment made on the Crockett furnace, and the results indicate that the success was mainly owing to the manner of firing, and to the fact the draft was controlled by the ash pit doors, in place of by the damper in the chimney, the effect being to make the area of egress larger than the area entrance for the air. The tan on the top of the furnace undoubtedly saved some heat. This experiment being the best performance of the Crockett, should be compared with the best performance of the Hoyt and Thompson furnaces at Wilcox and Binghamton. If the air in the chimney had not been diluted with 25 per cent. of air from the outside, its temperature would have been 650°.

EXPERIMENT MARKED WELLS' (4).

This experiment was made to determine the effect of reducing area of exit for gas. The rate of combustion was the same, the manner of firing was the same, the temperature of the furnace was nearly the same, and the chimney 10° colder.

Notwithstanding, the steam generated by one pound of dry tan was 5 per cent. less. The area of flues in this experiment was reduced until they had some influence on the gas flowing from the boiler, and resulted in a small loss of combustible gas. There was some smoke during this exeriment.

During the experiment marked Wells' (1) there was an excess of steam formed over that required to run tannery.

During the experiment marked Wells' (3) there was sufficient steam formed to carry 60 pounds pressure and run the whole tannerry, including heating liquors and all the machinery they ever ran. The consumption of tan was at the rate of 4 cords in 12 hours, or, allowing 1 cord for banking fires, 5 cords per day, being less than one-half of the bark they used for tanning their hides.

MANNER OF MAKING EXPERIMENTS.

The experiments were all made in the same general manner as follows:

Tan—The tan was measured in a box at its natural density; that is, merely shoveled into the box and "struck" with a straight edge. The box was counterbalanced on platform scales, and the weight of tan contained in each box noted in the log. After the furnace had been running a few hours, so that the amount of tan required was known, it was arranged to deliver a boxful every 5, 10, 15 or 20 minutes, according as required. The boxes, under these circumstances, were dumped at *exactly* the end of these intervals; that is, if there were required 3 boxes of tan per hour, one would be dumped at the even hour, one at 20 minutes past, and one at 20 minutes before, *exactly*. The object of this arrangement was to prevent any confusion as to the number of boxes delivered. When, as at Wilcox, there are used sometimes 20 boxes an hour, such an arrangement is found to be necessary. The tally of the weight of each box, and the minute at which it was dumped, was kept by myself and the man who was in charge of the gang of shovelers.

Water in tan—The water in the tan was determined by drying several specimens of 200 grammes each, taken from a sample of one quart brought from the tannery in a hermetically sealed glass jar. The tan was dried by my associate, Prof. J. K. Rees, of Columbia College, in an air bath at 110° Centigrade. The method of obtaining the sample of one quart was as follows: There was provided a tin case with a lid and lock. A double handful of tan was taken from each boxful of tan as weighed upon the scale and thrown into the tin case. The case was kept locked except when opened to receive samples. At the end of the day there would be collected in this case from 40 to 200 double handfuls of tan, *being collected in equal portions from every boxful of tan used.*

At the end of the day the tan in the case was shaken well up and then spread out upon a table and divided into equal small parts, alternate small parts being taken and the balance rejected. The original caseful was thus reduced by successive divisions (one half being rejected each time) to one quart, when it was carefully placed in the jar, sealed, labeled, and sent to New York by express.

Water—The water pumped into the boiler was all measured in casks, the weight contained by the casks, weighed on the same scale as the tan, having been originally determined. There were generally three tiers of casks, the lower one being connected with the pumps and acting as a reservoir, the middle one acting as a measuring cask, and the upper one receiving the supply of water and being provided with an overflow at the top and a plug in the bottom. The middle or measuring cask also had a hole in the bottom and a plug.

The mode of operation was as follows: The man in charge of the water would, when the measuring cask was empty, put the plug in the bottom and pull out the plug from the upper barrel, the water immediately commencing to flow from the upper into the measuring cask. When the measuring cask was full, he would put the plug in the upper barrel, and after a few seconds, when the surface of the water had subsided to the level of the overflow hole, would pull out the plug in the measuring barrel and allow the water to flow into the lower barrel, whence it would be pumped into the boilers. The same precaution was taken to avoid confusion by emptying barrels at exactly even intervals, as in weighing tan.

Water entrained with steam—This was determined by a special apparatus designed for the purpose and carried around to all the tanneries tested. It consisted of a worm, through which a small portion of the steam coming from the boiler was passed, exposed to cold water on the outside. The measurements taken were:

1. The weight of water coming from worm per hour being the sum of the steam and water coming from the boiler into the worm.

2. The weight of water passing over the outside of the worm, which carried away all the heat abstracted from the water and steam.

3. The temperature of the water before and after passing the worm, the temperature of the water delivered from the worm, and the pressure of the steam in the boiler.

These three measurements supply all the data necessary to determine the amount of water entrained with the steam.

The weight of water passing the outside of the worm was determined by noticing the head of water over an orifice in the bottom of the box in which the worm was, necessary to force all the water through. The *volume* of water which would be delivered by this orifice for each ¼ inch increment of head had been previously determined, and was as follows:

CUBIC FEET OF WATER DELIVERED PER HOUR FROM ORIFICE IN BOTTOM OF TANK AT VARIOUS HEADS FROM 16 INCHES TO 25 INCHES, AT A TEMPERATURE OF 72 DEGREES F.

Head.	Cubic feet water.	Head.	Cubic feet water.
25 inches	39.60	20¼ inches	35.47
24¾ inches	.22	20 inches	.26
24½ inches	.02	19¾ inches	.05
24¼ inches	38.82	19½ inches	34.82
24 inches	.62	19¼ inches	.59
23¾ inches	.42	19 inches	.36
23½ inches	.22	18¾ inches	.13
23¼ inches	.01	18½ inches	33.90
23 inches	37.80	18¼ inches	.68
22¾ inches	.60	18 inches	.45
22½ inches	.39	17¾ inches	.21
22¼ inches	.18	17½ inches	32.98
22 inches	36.98	17¼ inches	.74
21¾ inches	.77	17 inches	.50
21½ inches	.55	16¾ inches	.26
21¼ inches	.34	16½ inches	.02
21 inches	.13	16¼ inches	31.78
20¾ inches	35.91	16 inches	.53
20½ inches	.69		

The figure given in the log is the "head of water in tank." By entering this table with the head, the volume of water corresponding can be determined.*

*It is believed that the idea of determining the water entrained with the steam by condensing a small portion continuously in a worm and weighing and measuring all quantities, originated with Mr. J. D. Van Buren, Jr.

Temperatures—The temperatures, when below 600°, were taken by a thermometer. Above 600°, they had to be approximated to by means of the melting point of metals. Specimens of the same piece of metal were used at all furnaces, and the results are therefore fairly comparative.

Method of proceeding—The engineer generally took charge of the firing, and conducted it in the manner he supposed best calculated to produce the best result. As all of the engineers seemed to feel considerable pride in having their furnaces do well, I have reason to suppose each did his best.

Supervision—I was present at the furnace during the whole of each experiment, never leaving it for a single instant, and had a complete supervision of all operations during the whole time.

Ashes and refuse—These were determined by raking out the fire and commencing with the tan that had been weighed (in Stevens' and Wells' 1), and by weighing the ashes and refuse dry at the end of experiment. The ashes were found to be so small a percentage, and so much mixed with unconsumed tan, that the attempt to measure them was given up after these experiments.

In all others the fire was left at same condition at end as beginning, *i. e.*, Thompson and Wilcox furnaces *full*, and Crockett average thickness.

THOMPSON FURNACES.

These furnaces are built in some respects as described in the patents of Moses Thompson, as interpreted by the court. They consist of a series of ovens arranged in pairs, each pair of ovens with the boilers attached being complete in itself. The ovens are built of fire-brick, generally from 4 to 4½ feet in diameter and from 10 to 12 feet long. The crowns of the ovens are semicircular, and the grates are in the horizontal diameter.

The ovens are fed with wet tan through two openings in the top of each oven. These openings are about one foot in diameter, and so placed in the top of the oven as to best distribute the tan uniformly over the grate. They are supplied with iron covers, but except when starting or burning down the fires these covers are not used, as the tan, kept two feet deep all over the fire room floor, effectually seals these openings. The tan is fed as often as in the judgment of the fireman it is required.

When about to feed, he first breaks down the layers of bark which have dried and baked hard over the top and in the feed hole, and then with a bar spreads out the mass of embers and ashes left from the tan last fed in, and finally shovels into the oven all the bark that will run into the feed holes, and ends by filling all up and tramping down the bark into the hole, and covers the top with about two feet of wet tan as before.

So far as in the method of construction and operation, I do not understand that the representatives of Thompson make any claims to novelty.

Their peculiar claims are as follows :

The grates are made of fire brick and are $2\frac{3}{4}$ inches wide on top and have a space between them $\frac{1}{2}$ inch wide, the total area of openings between the bars for the admission of air being less than 20 per cent. of the total grate surface, while in the ordinary forms of grate for coal it is about 33 per cent.

The space between the bars ($\frac{1}{2}$ inch wide) allows a small amount of tan to run through, partly from its own weight, during the whole process of combustion, but mostly while the fire is being stirred preparatory to being fed. The whole amount that runs through is, however, a very inconsiderable proportion, and would scarcely be felt if it were lost. I do not think it can exceed 10 per cent. in either of the furnaces experimented on, and was probably much less. The amount that runs through, to a certain extent, regulates itself. It falls into a second oven, built of fire brick, taking the place of the ordinary ash pit. This oven is double for each furnace, the dividing wall being necessary to support the fire brick grates. In the furnace at Binghamton they were each 22 inches wide and 12 feet long. All the air that enters the oven through the grate passes through a small register in the ash pit door into the ash pit, and may then rise through the grate into the furnace, or may pass along under the grate over the glowing surface of the embers in the ash pit. The amount of air reaching the embers in the back end of the ash pit is necessarily small. The embers falling through the grate slowly accumulate in the ash pit, not consuming as fast as they pass through, until the top of the pile of embers touches the bottom of the grate bar and prevents the falling through of any more. During the time the embers are accumulating they only fall through very slowly, perhaps two or three a second.

When the embers are accumulated in the ash pit nearly up to the grate bars, the whole surface of the mass of embers, brick work and grate bars has a very high temperature, perhaps 2,000° F. The reason for this is, I conceive, not on account of the quantity of tan consumed there, or on account of the perfection of the combustion, but because the heat radiated from the embers is *not carried away.* Only a very small quantity of air circulates through the ash pit under these circumstances, and as all the ash pits in the group, except the two outside ones, are surrounded by others equally hot, but *little heat is lost by radiation.*

When, on the other hand, the surface of the embers is six or eight inches below the under side of the grate bars, the temperature of the gas at the back end, after passing over the whole length of the ash pit, is *not sufficient to melt lead.*

The other peculiarity of the Thompson furnace, as built at Binghamton or Brackneyville, is the contraction and combination of the outlets for the products of combustion from *each pair of ovens.*

The inventor claims that by contracting and combining the flues leading from the furnace to the boiler, any combustible gas coming from one furnace will be consumed by the excess of oxygen in the products of combustion coming from the other, or, in the words of Prof. Silliman, "the gases from the two furnaces may mingle and consume each other."

In order that this may take place these three conditions must simultaneously exist :

1. There must be combustible gas in the products of combustion of one furnace.

2. There must be an excess of oxygen in the products of combustion of the other.

3. There must be a sufficiently high temperature in the mixed gas (about 1,500°).

There was no provision for the admission of air into the combustion chamber of either the Thompson furnace at Binghamton or Brackneyville other than that which found its way through the oven without uniting with the incandescent tan—that is, that passed through an oven 10 feet long filled with tan at a temperature of nearly 2,000°. If there were any combustion taking place in this chamber I judge there must have been some appearance of flame there. There was a small hole drilled in the door of the chamber of the furnace at Binghamton (Weed's) which allowed an observation of the whole interior at any time without opening the door. The whole interior was nearly always perfectly clear. Occasionally a little flame came *out* of the oven *into* the chamber.

WEED'S FURNACE AT BINGHAMTON.

The experiment was made on this furnace by allowing the engineer to take entire charge of the method and quantity of firing, he proceeding as he said he had found by long experience to be the best way.

The tan was brought from the leaches and dumped on the fire room floor, where it was kept about two feet thick the whole time.

The fireman fired the oven without any apparent system, the average number of times each oven was fired being, including both holes, twice every 3 hours or 8 times during a day of 12 hours. Sometimes both feed holes were fed at once in one furnace of the pair, sometimes one feed hole in each was fed, and sometimes both feed holes in each furnace during the same hour, but generally one feed hole in one oven in one hour and the opposite feed hole in the next oven during the next hour, the feed holes nearest the boiler being fed every four hours, and those farthest from the boiler every two hours.

The ash pit doors were arranged so as to expose an area of nearly one-third square foot for each oven, or one-sixth square foot for each ash pit, and were not changed during the experiment. One ash pit was raked out partly, the other being nearly full of incandescent embers.

There was scarcely any smoke issuing at any time from the chimney, it being just perceptible occasionally.

The rate of combustion remained nearly constant during the whole day, and was nearly 65-100 pounds of dry tan per hour per square foot of heating surface.

The tan on being fed through the feed holes into the oven immediately commenced to give off vapor from its surface, and, as soon as the immediate surface was dried, to undergo a species of destructive distillation giving off combustible gas. This gas accumulated in the furnace apparently faster than it was carried away, for the smoke would, just after firing, return into the fire room through the tan packed into the upper part of the feed hole. The fire room was continually full of smoke. Part of this came from the combustion of the tan when heaped around the sheet iron work of the front connection, and part of it from the destructive distillation of the wet tan on the top of the ovens, the lower layers being found perfectly charred, but I am of the opinion that part came from the oven back into the fire room.

The gas from the furnace, after passing into the combustion chamber, passed forward under the boilers and returned through the tubes into a sheet iron front connection, thence into the stack. The boilers were unprotected on top and had steam drums.

The tan on being fed into the furnace immediately spread out into a cone, the angle at the vertex being nearly 90 degrees, reaching from the feed hole to the grate.

Just before feeding the surface of the oven would be a bright red all over, which would gradually cool down to a very dull red in that part of the oven nearest the feed hole last fed. The weight of the charge of wet tan fed in was nearly 300 pounds. The combustion seemed to take place entirely on the surface and to reduce the volume of the cone by equal decrements of volume from all parts of the surface, to a very obtuse cone about 60 inches in diameter and 24 inches high. During the process of combustion the oven would grow steadily hotter to the end, and would be, when fed again, a bright red heat, probably about 1,800°.

Silver exposed in the combustion chamber was readily melted at all times, sometimes, however, a little sooner than at others. The shortest time was about 30 seconds and the longest about 3 minutes required to melt a silver "dime." A piece of silver, however, hung on a chain and suspended in the furnace by passing the ends of the chain through the two feed holes, the position of the coin being about midway between the feed holes and between the grate bars and the crown of the oven, thus being exposed to the direct action of the flame in the furnace and of the radiant heat from the tan, was not melted, although exposed during the whole experiment. This must, I think, have been because at that point the products of combustion of the tan from one cone were diluted with the air which would afterward be partly consumed by the tan of the other cone on its way to the chimney. There was a perceptible variation of the color of the brick work of the arch, when the tan in both cones was consuming, from a bright red at the end nearest the boilers to a dull red at the end farthest from the boilers. I think a coin exposed in the oven between the second cone and the combustion chamber would have been melted as easily as in the combustion chamber, although this is only an opinion. The fact that the temperature was higher after passing the second cone than before shows that there was a greater excess of air in the products of combustion before passing the second cone than afterward, and suggests the idea that several successive cones of incandescent bark (which would be the result of four or more feed holes, as in the Hoyt furnace), might still further reduce the excess of air, and therefore increase the temperature of the gas.

The time during which distillation only is taking place, that is, before the surface of the cone becomes red, does not exceed five minutes, but it is probable that the tan is only dried and coked to a small depth during that interval, and that the continued distillation goes on from the outer layers of the cone from the commencement until near the end. On raking away the outer surface of the tan during the early part of the process, the fresh, green undried tan will be found underneath.

All brick and ironwork around passages for products of combustion from oven through combustion chamber were nearly air-tight. Only an insensible quantity of air can have leaked into the flues and mingled with the gas, en route, until the back connection is reached. This was of sheet iron and may have leaked considerable air, and also the front connection may have leaked some air.

The gases on reaching the back connection had sufficient temperature to heat a small spot on the cast iron doors to a dull red heat in the dark.

The amount of air found per pound of dry tan in the products of combustion was (by means of the velocity of smoke in the chimney) 10½ pounds. The relative amounts of carbonic acid and free oxygen in the products of combustion by volume, in per cent. of the *dry gas:*

Carbonic acid	11.1
Free oxygen	10.6

The heat developed was (see table), (13-100 being allowed for radiation per pound of dry tan), 7½ pounds of water requiring 6 pounds of air. If the air in excess is 10.6-11.1 of the carbonic acid, the air supplied per pound of dry tan would be 21.7÷11.1×6=11.7, or nearly a coincidence.

If this air had all passed through the combustion chamber, the temperature could only have been (calculated from the)

Elevation of temperature of 10¾ pounds of air, 1 pound of fuel and 1.22 pounds of steam, the air and fuel being supplied at 70 degrees and the steam at 212 degrees by the available heat in the pound of dry tan being 7½—1.44=6.06	1,800°
Elevation of the products of combustion being 10¾ pounds air, 1 pound fuel and 1.22 pounds steam from the temperature of the chimney 510 degrees by the addition of the heat abstracted by the steam and radiation being 4.43+.13=4.56	1,810°

The temperature of the furnace being sufficient to melt silver readily (2,000°), would seem to show that an amount of air equivalent to 10 per cent. of the products of combustion, or 1 4-10 pounds of air, had leaked in, leaving the pounds of air per pound of dry tan which passed through the furnace 9 1-10, or nearly the same in proportion to the heat developed as was found at Wilcox, or $1\frac{1}{2}$ times that actually necessary for perfect combustion. It must be acknowledged that this furnace was injured by 10 per cent. of cold air leaking into the flues between the combustion chamber and the chimney.

Notwithstanding this excess of air there is an amount of heat equivalent to 1 1-10 pounds of steam which was not developed by combustion, being about 12 per cent. of the heat in the tan or 25 per cent. of the heat realized on the steam in the boilers.

The cause of the incomplete combustion was probably partly due to the small supply of air, for it has been found that in order to burn all the combustible gas coming from bituminous coal in an ordinary furnace, the products of combustion must be diluted with 40 per cent. of their volume of air. In the case of the wet tan at Weed's furnace the products of combustion of the tan in air were already diluted with 40 per cent. of their volume of steam, and therefore in order that the proportion of consumed gas and the nitrogen and water diluting it should be the same as in the case of bituminous coal, the supply of air must be $1\frac{3}{4}$ that actually necessary for combustion.

There is a way in which the performance of the Thompson furnace is sometimes injured, and in which many of the earlier Thompson ovens that did not succeed were probably injured, that is, by a too great contraction of the flue leading from the oven to the combustion chamber. The effect is the same as on the Crockett furnace at Great Bend. The effect of the contraction at Binghamton will be considered further on.

These outlets were arranged as shown in the plans and had an area of 3 6-10 square feet for each pair of ovens, being 1-23 the grate surface. The cross section of the tubes in one boiler was 2 8-10 square feet. The temperature of the gas on leaving the furnace was nearly 2,000°, and at the entrance to the tubes 1,000°. The volumes of gas were then in the ratio of 5 to 3, and the area of the passages in the ratio of 4 to 3, indicating that the passage from the furnace to the combustion chamber might have been enlarged at least 1-4 with benefit to the boiler without reference to the combustion.

THOMPSON FURNACE AT BRACKNEYVILLE.

This furnace was similar to the one at Binghamton except that there were two flue boilers, each 3 feet 9 in diameter and 20 feet long, with two flues in each boiler 12 inches in diameter. Only 3 of the 4 ovens were in use, the fourth being bricked off by a brick wall in the opening into the combustion chamber. I had no means of knowing how nearly air-tight this wall was.

The average rate of combustion was 73-100 pounds of dry tan per square foot per hour, but as in one boiler it was 97-100 and in the other 49-100, the average does not measure the relative economy.

Silver could be melted in the combustion chamber (the one connected to the pair of ovens) but *could not be introduced into the oven.*

The performance of this furnace was in every respect similar to the one at Binghamton, except that as the rate of combustion was only one-half as much per square foot of grate the ovens were only fed half as often. There was no system of alternate feeding, and no changing of the ash pit doors.

The furnace was not in as good repair as at Binghamton.

The pounds of air per pound of dry tan were found to be 18, by the velocity of the smoke in the chimney, and by the analysis of gas as before, 17 9-10 pounds.

Assuming the 18 pounds of air as nearest correct we find for the temperature of the furnace, computed as before :

From thermal equivalent of tan	1,320°
From temperature of chimney and steam	1,350°
	1,335°

From which may be computed the proportion the air found in the chimney bears to the air passing through the furnace, in order that the combustion chamber may have a temperature of 1,800°, and the weight of air that passed through furnace :

Pounds of air per pound dry tan	12
Ratio of air necessary to air supplied	100 :175

The weight of air leaking into the flues between the oven and the chimney was 6 pounds per pound dry tan.

HOYT FURNACE AT WILCOX.

This furnace consists of two pairs of ovens, each pair of ovens being connected with 3 horizontal flue boilers. Each oven, with its boilers, is entirely independent of the other, each being furnished with a separate feed pipe, steam pipe, water tank, safety valve and injector. The only things in common between the two ovens are the fire-room and the chimney.

The chimney is of brick, 107 feet high above the grate.

The ovens are each 6 feet wide and 16 feet long.

The grates are of iron, the width of the bar being 7-16 inches and of the space 7-32 inches.

The ash pits are the entire width of the grate and entirely open in front, no doors being provided.

The distance from the under side of the grate bars to the bottom of the ash pit is 4 feet 9 inches. This great height allows a double current of air to form in the ash pit, the cold one entering at the front near the bottom, passing toward the back end, becoming gradually warmed by the radiant heat from the grate, rising and returning close under the grate, part entering through the grate to the oven, and the balance finally passing out of the ash pit at the front at a temperature of 300°. Under these circumstances the temperature of the ash pit is about 200°.

If the furnace is so far closed as to prevent the return current of air, that is, if the opening is reduced from 24 square feet to ½ of 1 square foot, the temperature of the ash pit rises to about that of the Crockett furnace, 500°.

When the front is open there is a small loss, due to heating the air which returns to about 300°.

There is no contraction of the flue from the oven to the space under the boilers, the opening being the full width of the oven.

There are four feed holes, each one foot in diameter, in each oven. The tan is fed in through these holes as it burns down, and the holes sealed by wet tan kept all over the floor about two feet deep.

The bark is brought about 250 feet by a conveyor from the leaches, and may be delivered at any point on the floor by opening a suitable shoot.

This furnace was the most complete in all its appurtenances, being of originally the best design and in the best condition of any experimented upon. The boilers were new and tight, the brick work sound, and all parts of the boiler and smoke connection protected from loss of heat by radiation. The chimney was of brick.

There were dampers in the flues leading from the boilers to the chimney. The cross-section of these flues was 4 square feet each. In the case where the most tan was burned the damper was kept wide open, and in the case where the least was burned it was kept at an angle of 45° in the pipe, the effect being to reduce the area to about 2 square feet.

The pounds of dry tan burned per hour were respectively 1,000 and 750, or a proportion of 6-10 and 4-10 square inches to one pound of dry tan per hour. This opening was not the controlling section in the first case, although it was in the second.*

The performance of the furnace during the first experiment does not correspond with the *maximum*, which might be 25 per cent. more, or 1,250 pounds of dry tan per hour, or ¾ of a cord. The draught was checked in this experiment by a thick bed of ashes which had been accumulating on the grate without being raked out for 10 days. During this time the ashes had accumulated 8 inches to 12 inches deep. The other furnace had not been raked out for 4 days.

No ashes or embers *fell through* the grate at any time. The ash pits were raked out in expectation that some would collect, but were found perfectly clean at the end of the experiment. It is customary to rake out the furnace through the 8 doors provided every fortnight. If it were required to drive the furnace to its utmost it would be necessary to rake them out every week *at least*.

The action, manner of feeding, and appearance of the Hoyt furnace in both experiments is very similar to the Thompson furnace already described, except that the temperature of the oven was more nearly uniform, and there never was any smoke forced back into the fire room through the feed holes.

The fires were watched through the small registers provided in each furnace door, and were filled up when burned down. They were generally allowed to burn down lower, *i. e.*, further from the crown of the furnace than at Binghamton or Brackneyville, the greater height of the furnace, 6 feet in place of 4½ feet, allowing this.

In this connection it must be remembered that while the rate of combustion per square foot of grate surface was at Weed's the same as at Wilcox (2), and therefore the rate of combustion per square foot of superficial area of the cones of tan the same, the actual weight of tan in the furnace undergoing the processes of drying and coking was nearly 2½ times as much at Wilcox as at Weed's, and therefore the performance of the furnaces, so far as all phenomena regarding the intermitted action of the drying and coking are concerned, are not comparative.

There was no smoke coming from the chimney at any time during the experiments.

*The openings of 6-10 and 4-10 square inches for one pound of dry tan burned per hour are equivalent to 8-10 and 6-10 square inches, with a chimney 80 feet high, and having a temperature of 600 degrees inside and 60 degrees outside, as in the case of Wells'.

A silver coin exposed in the back end of the oven melted nearly as soon as at Weed's, in about 60 seconds.

Much of the air for combustion came through the registers in the furnace doors, for although these registers, except for a few seconds when looking at the fire, were kept shut, they were only iron casting, and must have leaked considerably through the badly fitting joints. I do not consider that the air leaking in here did any harm, but acted as efficiently as if it had come through the grate.

When these registers were opened to examine the fire, the surface of the heaps of tan appeared to be covered with about equal portions of black and red cinders, and would begin to scintillate and grow rapidly red when the air from the register touched it, showing that at that part of the furnace at least, the oxygen in the air surrounding the heaps of tan was reduced below the point of dilution necessary to insure a rapid combustion.

A considerable amount of heat was lost by radiation from the oven doors. There were 8 of these doors to each furnace, and they all became a dull red in the dark over about 1 square foot of their surface, the atmosphere outside being 30°.

The labor of firing at this furnace was very small, owing to the convenient arrangement for delivering the tan on the fire room floor. One man fired usually both furnaces and tended to the water during the day, thus handling usually 10 cords, or 40,000 pounds, of wet tan in 12 hours.

If the temperature of these ovens as actually measured be compared with the calculated temperature by the two methods as before, taking the air at 10 pounds per pound of dry tan, we have the following:

	Wilcox (1)	Wilcox (2)
Temperature of oven computed from the known weight of air and the thermal equivalent of the tan	2,060°	2,060°
Temperature of oven computed from the heat given up by products of combustion to steam and adding temperature of chimney	2,050°	2,080°
	2,055°	2,070°
Average		2,060°

This temperature is a little in excess of that estimated from the melting point of silver, showing that no air can have leaked into the flues between the oven and the chimney.

The air for combustion was then 10 pounds per pound of dry tan, being 1 40-100, that actually necessary for combustion, or 40 per cent. of "air in excess," being less than at any other place where the combustion was perfect. I think this may be due to the long oven and 4 cones of tan, by means of which the air in the furnace is intermingled, and each particle brought in contact with surface of tan.

LEACHED BARK.

Wood (and inferentially bark) is said by M. Violet to consist of a fibrous substance called "cellulose," which is insoluble in water, surrounded by a material which is richer in carbon and hydrogen. This encrusting material consists mainly of sugar, starch, gum, resin, glucose and tannin. All of these, except the resin, are soluble in hot or cold water, and are probably wholly or partially extracted in the process of leaching.

They are all compounds of hydrogen, oxygen and carbon, but being rich in oxygen will develop by combustion but a small amount of heat. On the other hand, the resin, which is perfectly insoluble in water, and must remain entire in the bark after the most thorough leaching, contains no oxygen, and has a thermal equivalent of nearly 22 pounds of water evaporated from 212°, or about three times as much as the soluble constituents.

It would appear then that the thermal equivalent of leached bark must be more than of green or unleached bark.

The average thermal equivalent of the soluble constituents of the bark is 6½ pounds of water evaporated from 212°, and of the unleached bark (as has been already determined), 8½ pounds of water evaporated from 212°.

There will be extracted from a cord of unleached bark in the process of leaching, if thoroughly done, 600 pounds of soluble matter, and the remainder will weigh 1,400 pounds when dried at 110° C.

If the thermal equivalent of a pound of the bark as leached at the tanneries in the different experiments, be calculated on the basis that the original equivalent was 8½ pounds, and that a cord of leached bark has lost the balance of its actual weight, and 2,000 pounds of soluble material having an equivalent of 6½ pounds, we shall find for the thermal equivalent of each bark in pounds of water evaporated from 212° by one pound of bark dried at 110° C.:

Stevens'	8.9	Wells' (2)	9.0
Wells' (1)	9.0	Wilcox (1) and (2)	9.0
Brackneyville	8.9	Wells' (3)	8.9
Weed	8.6	Wells' (4)	8.9

This would be the weight of water evaporated or the weight of steam formed in the boiler by the combustion of one pound of tan if the combustion in the oven were perfect, and if the boiler surface were sufficiently extended to reduce the products of combustion to the same temperature as the outside air, or about 60°.

It will be seen that only about one-half of this amount was evaporated in any case, and generally less than one-half.

There are four principal reasons for this discrepancy :

1. Water in the bark.
2. Imperfect combustion.
3. Radiation.
4. Heat carried away in the chimney gas.

The first of these is entirely independent of the oven or boilers, and if the proportion of water to bark were the same, might be neglected in every case without affecting the *comparison*. Unfortunately the proportion of water varies in every case. The water put into the oven with the bark has to be evaporated, and absorbs the same amount of heat in being evaporated as though it were put into the boiler.

It also lowers the temperature of the oven, thus injuring the effect of the heating surface.

It also dilutes the air entering the furnace, so that probably the wetter bark will require a larger excess of air for equally perfect combustion.

The heat in a pound of dry bark that is *available*, then, is the total heat given above reduced by the heat necessary to evaporate the water in the bark, and to superheat it to the temperature of the chimney (see table) and will be :

Stevens'	6.9	Wilcox (1)	6.8
Wells'(1)	6.9	Wilcox (2)	6.9
Brackneyville	7.0	Wells' (3)	6.9
Weed	7.0	Wells' (4)	6.9
Wells' (2)	7.0		

This is the weight of water that *might be* evaporated from the boiler by each pound of dry tan.*

INCOMPLETE COMBUSTION.

Generally in all cases a portion of the heat, about 5 per cent., is lost by incomplete combustion. No analysis of the chimney gas has been made in these experiments that will show how perfect the combustion was. It can only be inferred from the amount of heat accounted for in all other ways, probably being

*The injurious effect of a small proportion of water in fuel is much less than is generally supposed. A mixture of 100 parts, containing 13 parts of wood and 87 parts of water, will develop sufficient sensible heat to raise the products of combustion to the temperature necessary for ignition, about 500 degrees. The same is true of a mixture containing 7 per cent. of charcoal and 93 per cent. of water.

at least 2½ per cent. in the best case (Hoyt's furnace at Wilcox), and much more in the worst.

RADIATION.

This loss depends upon the form, arrangement and dimensions of the furnace and boiler, the absolute amount being in general proportioned to the surface of brick or stone work exposed. All the furnaces must have suffered considerably from radiation, but if we neglect the heat radiated by the brick work over the oven, which is utilized in partly drying and heating the tan in all furnaces fed from the top, all the furnaces were injured probably nearly alike, Wells' 2, 3 and 4 being injured most by radiation from brick work, and Weed most from radiation from sheet iron work in back connection, and Wilcox by radiation from furnace doors.*

CHIMNEY GAS.

The heat carried away in the chimney gas amounts to 30 per cent. in Wells' (1) and 15 per cent in Weed's.

It depends upon two items, viz.:

1. The excess of temperature of the chimney gas over the outside air.

2. The weight of the gas in comparison with the weight of the fuel.

The nearer the temperature of the gas is to the temperature of the atmosphere, the less heat will be carried off in the gas. It is the duty of the boilers to reduce the gas to the temperature of the chimney from the temperature of the furnace. *If the chimnies have the same temperature the boilers may be considered equally efficient, whatever may be the extent or arrangement of their surface*, provided no heat is lost by radiation, and no cold air leaks into the flues from the outside.

All the boilers tested were injured by a certain amount of air which leaked into the flues from the outside, generally through the fissures in the brick work or through the joints in the sheet iron work. Probably Stevens' and Wells' suffered most from this cause. At these furnaces the stone and brick work was old and full of fissures. During the many years they had been in use the alternate contraction and expansion of the walls had forced out the mortar from the joints. During some stages in the combustion *smoke issued* from these fissures, showing that there were open-

*The surfaces exposed for radiation were (iron work):
Weed, surface bark connection....50 square feet gas inside....1,000° outside....80°
Wilcox, surface furnace doors.....50 square feet gas inside....2,000° outside.... 80

ings through the wall, and that the air must flow in when the pressures were reversed.

Brackneyville was nearly tight in the brick work, as were Weed and Wilcox, but all these were injured by air leaking in through the sheet iron work around the back connection.

I have no means of estimating the amount of air leaking in through the flues, but should consider that it may have been insensibly small at Wilcox.

In addition to the air leaking into the flues some air leaked into the chimney between the boiler and the thermometer. This air had no effect on the economy of the boiler, but only reduced the temperature of the chimney and injured the draught. The temperature of the chimney, as measured by the thermometer, cannot be taken as a measure of the efficiency of the boiler surface. In Wells' and Stevens' in particular there were large cracks around the junction of the smoke stack with the boiler.

The leakage of air into the chimney had no effect upon the amount of heat carried away by the chimney gas, the temperature being lowered as fast as the weight was increased.

WATER ENTRAINED WITH STEAM.

The measurements taken to determine the water in the steam are given in the appendix. The results are as follows:

	Per cent. of water in steam.	Pounds per hour.
Wells' (1)	1.4	33
Brackneyville	7.9	95
Weed	3.4	168
Wells' (2)	3.5	55
Wilcox, superheated		24°

The close approximation of the absolute amount of water carried from the boiler in the two experiments at Wells' suggests the idea that this water was due to some condensation of the steam after it had left the water in the boiler as *dry steam.*

There was considerable surface exposed to the cold air in every case, being generally the surface of the steam drums and pipes, and in some cases the tops and ends of the boilers.

The amount of surface in each case can be approximately computed from the plans accompanying this report, and was as follows:

Wells'	40 square feet
Brackneyville	130 square feet
Weed's	250 square feet
Wilcox (to atmosphere at 70 degrees)	60 square feet
Wilcox (to gas in furnace at 2,000 degrees)	50 square feet

And therefore the pounds of water entrained with the steam for each square foot of surface exposed per hour:

Wells' (2)	1.325
Wells' (1)	0.825
Brackneyville	0.731
Weed's	0.752

A perfectly clean steam pipe filled with steam at 60 pounds pressure, and exposed to the *still air* at 80° temperature will condense steam at the rate of three-fourths of a pound per square foot per hour.

The close agreement of the weight of water entrained with the steam, and the weight of water that must have been condensed by the cold air on the outside of the pipe, leaves no escape from the conclusion that *the water in the steam was entirely due to condensation in every case.*

As far as the ill effect of a certain proportion of water on the steam when used in an engine for heating liquor or other purposes is concerned, it can make no difference whether the water has been condensed from the steam or whether it is carried from the boiler as *water*, without ever having been steam at all; but as far as the heat abstracted from the fuel by the steam is concerned, it does make a very important difference, the heat abstracted from the fuel by the steam being the same whether any of it or all of it is afterward condensed, reaching the engine as all steam, partly steam and partly water, or all water.

The heat abstracted from the fuel in the last case being that required to evaporate all the feed water pumped in.

It is therefore necessary in such experiments to pay particular attention to the amount of surface exposed to the cold air.

TEMPERATURE OF CHIMNEY.

The temperature of the chimney as shown by thermometer and the temperature of the atmosphere, the difference being one element of the portion of heat lost, were:

	Steven's	Wells' (1)	Brack.	Weed	Wells' (2)	Wilcox (1)	Wilcox (2)	Wells' (3)	Wells' (4)
Temperature of chimney gas	700°	580°	580°	510°	420°	700°	580°	490°	480°
Temperature of atmosphere	50°	70°	50°	65°	50°	70°	70°	30°	30°
Difference	650°	510°	530°	445°	370°	630°	510°	460°	450°

PRIMING.

The priming was therefore nothing in all of the boilers experimented on. This agrees with all reliable experiments I have ever seen, the water primed as shown by the experiment being less than the instrument could be relied upon to detect. The weight of water pumped in is therefore a measure of the steam formed.

ABSOLUTE EFFICIENCY.

The absolute efficiency of all the furnaces, as indicated by the experiment, being the heat utilized in the steam compared with the *available* heat in the tan, or the proportion that the steam actually evaporated bears to the steam which might be evaporated by the same tan in a perfect furnace and boiler, no allowance being made for imperfect combustion, insufficient heating surface or radiation, are as follows:

Stevens'	33.6	Wilcox (1)	68.9
Wells' (1)	44.5	Wilcox (2)	75.7
Brackneyville	57.1	Wells' (3)	61.1
Weed	64.7	Wells' (4)	57.3
Wells' (2)	45.8		

The only furnaces that are fairly comparable are those at the same rate of combustion per square foot of boiler surface. The rates of combustion—that is, the pounds of dry tan burned per square foot per hour, were:

Stevens'	1.61	Wilcox (1)	1.05
Wells' (1)	0.98	Wilcox (2)	0.79
Brackneyville	0.73	Wells' (3)	0.63
Weed	0.67	Wells' (4)	0.63
Wells' (2)	0.64		

Of these experiments Wells' (1) and Wilcox (1) are at nearly the same rate; also Brackneyville and Wilcox (2); also Weed and Wells' (3); Wells' (2) and (4) and Stevens' being out as experiments made under exceptional conditions.

Comparing the experiments together, then, that have the same rate of combustion, there results:

	Hoyt.	Crockett.
Pounds dry tan burned per square foot of heating surface per hour	1.05	0.98
Absolute efficiency	0.689	0.445
Relative efficiency	100.	64.

For the next lowest rate of combustion:

	Hoyt.	Thompson.
Pounds of dry tan burned per square foot of heating surface per hour	0.79	0.73
Absolute efficiency	0.757	0.571
Relative efficiency	100.	76.

For the lowest rate of combustion:

	Thompson.	Crockett.
Pounds dry tan burned per square foot of heating surface per hour	0.67	0.63
Absolute efficiency	0.647	0.611
Relative efficiency	106.	100.

Although the experiments at Brackneyville and Wells' (1) cannot be compared directly together, they are both compared with the same furnace at different rates of combustion, and the resulting ratios may be compared (that is, 76 and 64), the final ratio being 100 and 118. The final ratios then are (each furnace being at the same rate of combustion):

Hoyt	122
Thompson	111
Crockett	100

These are the actual experimental comparative efficiencies of two Thompson, two Hoyt and two Crockett furnaces, furnace being taken in the actual condition found.

The question now arises, "How much of this difference is due to the principle of construction and how much to the superiorities or defects of each *particular furnace.*"

The answer to this question may be very accurately inferred from the figures already given in the report.

The main causes of difference are:

1. Incomplete combustion.
2. Dilution of air either passing through fire or leaking into flues.

INCOMPLETE COMBUSTION.

The table at the end shows the total heat accounted for in the six cases where the air passing up the chimney was measured, as follows:

	Wells (1).	Brack.	Weed.	Wilcox (1).	Wilcox (2).	Wells (3).
Steam	3.03	4.01	4.53	4.61	5.30	4.17
Water in bark	2.28	1.91	1.61	2.19	2.09	2.07
Gas	2.94	2.40	1.33	1.73	1.40	1.96
Water of combustion	0.45	0.45	0.50	0.45	0.45	0.45
Sum of above four quantities	8.70	8.77	7.97	8.98	9.24	8.65
Calculated thermal equivalent of bark	9.00	8.90	8.60	9.00	9.00	8.90
Difference being heat unaccounted for from all causes	0.30	—0.13	—0 63	+0.02	+0.24	—0.25

It appears from the above that the combustion was sensibly complete in all except Brackneyville and Weed (the discrepancies between Wilcox (1) and (2) showing the limit of accuracy of experiment).

INCOMPLETE COMBUSTION CAUSED BY CONTRACTED FLUE.

There is a way in which the contracted opening from the oven into the flue may act injuriously in the Thompson furnace if carried to excess, and is positively a cause for the failure of many of the earlier Thompson ovens, and for the incomplete combustion of the Weed furnace.

If the opening from the oven into the combustion chamber is as much contracted that it is the controlling section, that is to say, if it is so small that the principal resistance offered to the flow of the gas from the ash pit door to the chimney top is at this point, then a considerable portion of the combustible gas will escape unburned.

The cord of wet tan as fed into the furnace contains:

Water	2,700 pounds.
Gas	870 pounds.
Fixed carbon	600 pounds.
Ashes	30 pounds.

If completely burned the products of combustion will be at the temperature of the furnace (say 1,800°) and will have the following volume:

Steam	240,000 cubic feet.
Gas and steam and air	300,000 cubic feet.
Carbonic acid and air	600,000 cubic feet.
	1,140,000 cubic feet.

It appears, then, that there must pass through the outlet of the furnace 1,140,000 cubic feet for every cord of tan burned.

At the Weed furnace there was burned in each oven one-eight of a cord per hour, and therefore there must have passed through each outlet 140,000 cubic feet per hour.

If the outlet had been the controlling section, it would have passed this volume, or 2,300 feet per minute and *no more.*

The ovens were fed nearly every hour, and therefore the weight of the charge must have been $\frac{1}{8}(4,200)=505$ pounds, containing a volume of steam and gas that will be generated soon after its entrance into the furnace, equal to:

Steam	30,000 cubic feet.
Combustible gas	10,000 cubic feet.
	40,000 cubic feet.

being a volume sufficient to fill all outlet from the furnace and completely arrest the inflow of air, during 17 minutes, if the gas and steam were completely driven off in this time.

The loss in this case would be the heat equivalent to all the combustible gas, or about 25 per cent.

The process of completely drying and coking the tan lasts during the whole time, from one time of feeding until another, but is most rapid during the first ten minutes, and probably nearly all the gas generated during the first ten minutes in the Weed furnace was lost. The actual loss of heat from incomplete combustion was about 8 per cent., indicating that one-third of the gas passed away unconsumed in *this way*.

The fact that there was a pressu e, or at all events an equilibrium of the gas in the furnace over the atmosphere, just after firing, was shown by the fact that the smoke from the tan in the feed holes returned into the fire room at this time, and kept it so full of smoke as to be almost unendurable.

This was not so at Brackneyville or Wilcox.

In order that the outlet from the Weed oven should be the controlling section it must have an area of less than 3 square feet, for the opening into the ash pit that only passed air at 60° was ½ one square foot.

The actual area may be computed from Wells' (4), when the outlet was reduced until it was the controlling section, and must be, in order to be the controlling section, less than 400 square inches. The actual area of the outlet was 256 square inches, being too small to pass the gas and steam as fast as generated, and the air necessary for the combustion.

EXPERIMENT ON REDUCTION OF OUTLET.

An experiment was made marked Wells' (4), in which the outlet from the flues was reduced until it began to sensibly affect the draught. The area of the outlet was reduced to 2½ square feet, and the tan consumed ⅓ of a cord per hour, containing (at the temperature of the chimney):

Steam	100,000 cubic feet per cord
Gas and air	120,000 cubic feet per cord
Carbonic acid and air	240,000 cubic feet per cord
	460,000 cubic feet per cord

being nearly 2,300 cubic feet per minute. The furnace was fed every 15 minutes, and, therefore, the charge, 1-12 of one cord, containing:

Steam	8,300 cubic feet
Combustible gas	3,000 cubic feet
	11,300 cubic feet

or enough to completely fill all passages for egress from furnace for 4½ minutes. If the gas were all driven from the tan during this time all passages for egress would be filled, no air could enter, and all combustible gas must be wasted, and the loss would be, as before, about 25 per cent.

If the period of drying of tan extended from time of feeding until the time of feeding again, and the gas and steam were expelled uniformly, the loss would be zero.

I judge from the appearance of the fire that the time occupied by the gas and steam in perfectly leaving the tan was about 10 minutes, and that therefore the loss was less than 10 per cent. The actual loss from experiment was about 6 per cent.

The controlling section was found to be at the ash pit 100 square inches, and the section that commenced to affect the draft at the chimney 360 square inches, from which I infer that in order to perfectly consume the tan in a furnace fed from the front, or with two feed holes, the areas of opening for air must be, per pound of dry tan burned per hour (with a chimney 80 feet high and a temperature of 500°):

Ash pit	0.20 square inches
Outlet of furnace	2.50 square inches
Entrance to chimney	1.00 square inches
Entrance to flues (when one oven is connected to boiler)	1.50 square inches

and that an area of the following area per pound of dry tan per hour will be the controlling section, and will limit the rate of combustion:

Ash pit	0.20 square inches
Outlet of furnace	0.90 square inches
Entrance to chimney	0.50 square inches
Entrance to flues	0.60 square inches

In the old Thompson furnace at Wells' there were 4 openings from the oven into the combustion chamber, each said to be 18 inches square—1,300 square inches. They used to burn 600 pounds of tan per hour *with difficulty*. This explains why they had so much trouble to keep steam. They need all the steam coming from 500 pounds of tan per hour, and if they waste all the combustible gas they must burn 1,000 per hour. This would be at the rate of 1 3-10 square inch of outlet for every pound of dry tan, or nearly the *limit* of *the furnace* in good weather under favorable circumstances. If for any reason the maximum performance is suspended for a few minutes the tannery must stop and wait for steam. This they often had to do. Probably their whole difficulty would be overcome by enlarging the outlets from the furnace to 2½ square feet each. I have no doubt this may be done to advantage in many existing Thompson furnaces.

At Brackneyville the area of outlet from furnace was 1½ square inches per pound of dry tan per hour, being too small. The loss by combustible gas here was 5 per cent. of heat utilized.

The economy of Weed's furnace would probably be increased

25 per cent. by either doubling the area of outlet from the oven or by reducing the consumption.

The rapidity of drying and coking the tan in the furnace depends upon the surface of the tan exposed to the radiant heat (being equal to the grate surface in a furnace fed from the front, and to the entire surface of the cones in the furnace fed from the top), while the amount of water to be given off depends upon the rate of combustion, and the portion of the time occupied in the drying of the tan upon the thickness of the bed on the grate. Thus the water and gas will be given off more uniformly during the whole process of combustion if the fire is carried on the grate 12 inches thick than 8 inches, and if the hight of the cone in a furnace fed from the top is 6 feet than 4½ feet. The hight of the cone being limited by the size of the oven, it follows that for an oven 6 feet in diameter the areas of outlet may be smaller. It is for this reason I judge that the Hoyt furnace at Wilcox was able to burn at the rate of one pound of dry tan to 1.0 square inch of flue area, and to 6-10 square inch of chimney area, both being less than the limiting areas for a furnace 4½ feet in diameter, or a Crockett furnace, without loss, although it will be noticed that the heat accounted for in the second experiment is more than in the first, indicating better combustion at the slower rate. Also when there are four feed holes there is a greater chance of one cone being freshly fed and giving off steam and gas all the time. If there were cones enough the cross areas of outlet, chimney and flue might be reduced to the limiting areas without loss. I think that sufficient air for combustion of the gases would enter the Thompson oven in the portion of the rectangular grate left uncovered by the circular bases of the cones, for these cones are spaced further apart in the Thompson furnace in proportion to the hight than in the Hoyt. The portion of the grate left uncovered, or covered very thin, would remain or soon become bare. These openings fulfill the same part as the holes in the "Crockett" fire.

The furnace at Binghamton was arranged so that the ovens were all connected in pairs, the furnace at Brackneyville so that one pair was connected and one single, and the furnace at Wilcox so that each oven was single. The combustion at Wilcox was sensibly perfect, while at Binghamton it was very imperfect. There is no escape from the conclusion that connecting the furnaces together was at least of *no benefit*. The same may be said of the hot ash pit.

It appears from the figures given in report that the combustion was more perfect at Brackneyville, and that also the supply of air per pound of tan was most:

	Weed.	Brackneyville.
Percentage of total heat in bark lost by imperfect combustion	9.0	2.4
Pounds of air supplied per pound of dry tan	9.1	12.
Per cent. of air in excess of that necessary for combustion	50.	75.

It would appear from these figures that it was necessary to supply 1¾ as much air to the Thompson furnace as was necessary for combustion, and only 1 4-10 times as much as was necessary for combustion to the Hoyt. I believe this is actually the case, not because of any of Thompson's patents, but because in the Hoyt furnace there are four feed cones, and the air and combustible gas have a better opportunity to be thoroughly intermingled than in either of the Thompson furnaces when there were only two feed cones; also because a greater part of the air in the Hoyt furnace is admitted directly into the oven above the fire, and has an opportunity to mix with the gas and consume it, while in the Thompson furnace it must *first* pass through the fire and then consume the gas. It will be very hard to get a sufficient supply of air to pass through the fire without being taken up by the red-hot embers. I doubt very much whether it would be possible to have an excess of 75-100 of the air taken up in the oven of the Thompson furnace as built at Binghamton—that is, without some means of supplying fresh air into the combustion chamber. At Brackneyville air was supplied to the combustion chamber through leaks in the badly-fitting door. A few holes in the door of the combustion chamber at Weed's would supply the air necessary.

Assuming, then, that there will be required in a Thompson furnace with two feed holes 1¾ as much air as is necessary, or 12 3-10 pounds per pound of dry tan, and that the air supply in the Crockett furnace must be 1 6-10 times as much as in the Hoyt, or 16 pounds of air per pound of dry tan, the maximum absolute officiency of the three furnaces will be when the chimney gas is at 500°, and when the wet tan contains 62 per cent. of water:

Hoyt	82.6	110.8
Thompson	79.2	106.2
Crockett	74.6	100.0

In this comparison each furnace has been taken with the air supply *actually found by experiment*, while from what has been said it appears that the Crockett furnace will only *need* 13 pounds of air per pound of dry tan. I have not made a comparison on the basis of 13 pounds of air for the Crockett furnace because it was not an *experimental determination*, and may be open to criticism.

Nevertheless, I have no doubt that if each furnace were proportioned in the best manner, and the walls tight, there would only be required 13 pounds of air per pound of dry tan, when the relative efficiencies would be :

Hoyt	105.4
Thompson	101.2
Crockett	100.0

The best and an almost perfect result, has been obtained in a furnace having an open backed single oven, with a cool ash pit. The results of these experiments go to show that two of Thompson's claims, namely : the hot ash pit and the reduced outlet from the furnaces are absolutely injurious, and that the third, *i. e.*, the combination of a pair of ovens, is at most of no value, for a practically perfect result can be obtained without the combination *in one oven.*

It has been customary to compare experiments on these furnaces, by adding the water evaporated in the boilers to the water in the tan, and by dividing the sum for one furnace by the sum for the other, paying no attention to the size of the boilers, the method of firing, the state of repair, or to anything but the water evaporated and the tan burned.

The results thus obtained give no idea of the relative merits of the two furnaces, but as this method has been persistently adhered to in former reports, such a comparison is made here, with the following results :

The average of all of each kind	Thompson	109
	Crockett	110
The best furnace of each kind	Thompson	99
	Crockett	100
The poorest Crockett with the best Thompson	Thompson	137
	Crockett	100
The poorest Thompson with the best Crockett	Thompson	97
	Crockett	100

It appears that by comparing in this way and selecting examples, any result from a superiority of 37 per cent. to an inferiority of 3 per cent. may be obtained for the Thompson, while the relative efficiency of the two furnaces as *shown by the experiment is :*

Thompson	106
Crockett	100

The results of these experiments indicate that for a furnace having a chimney 80 feet high, and discharging the gas at a temperature of 600°, the atmosphere being 60°, and the tan being well leached and containing 62 per cent. of water, compared with the residue dried at 110° C., as follows :

CALORIFIC POWER.

One pound of wet tan will, when perfectly consumed, evaporate 2 pounds of water from 212°, or 5 pounds from 212° per pound of dry tan, being about ¾ of the water that would be evaporated by one pound of the dry portion or ordinary pine wood.

POOREST PERFORMANCE.

One-third of the heat may be wasted by careless firing or inferior design of oven, and that there will be evaporated under these circumstances 2½ pounds of water for each pound of the dry tan.

CROSS SECTION OF FLUES.

That in order to insure perfect combustion, the area of egress for the gas from the boiler must exceed that necessary for coal or wood, and must be for an oven with two feed holes, or fed from the front at least, for each pound of dry tan burned per hour at the

Outlet of oven	2.50 square inches
Entrance to flues	1.50 square inches
Entrance to chimney	1.00 square inches

and that the draught *must be controlled at the ash pit door.*

When two ovens are connected to one boiler, or when one oven has four feed holes, the areas of egress per pound of dry tan per hour may be :

Entrance to flues	1.00
Entrance to chimney	0.75

LEAST SECTION OF FLUES, ETC.

When the fire is forced to the uttermost, the number of inches of sectional area at the various points in the passage of the gas from the ash pit to the chimney top will be :

Entrance to ash pit	0.20
Outlet to furnace	0.90
Entrance to flues	0.60
Entrance to chimney	0.50

COMBUSTION PER SQUARE FOOT OF GRATE.

Under the above circumstances the tan will burn at the rate of 10½ pounds of dry tan per square foot of grate per hour at least, and possibly more. It is, however, favorable to economy to have a slow combustion per square foot of grate surface, not exceeding

5 pounds per square foot of grate per hour, and probably even slower.

HEATING SURFACE.

There will be required in the Crockett furnace when burning 5 pounds of dry tan per square foot of grate surface per hour 1½ feet, and in the Hoyt furnace 1¼ feet of heating surface for each pound of dry tan burned per hour, being in the last case about the same proportion as would be required for a furnace burning coal or wood to furnish about the same amount of heat.

Although the heating surface required for coal or wood is the same as for tan, the cross section of the flues is required to be about double, and the grate surface about four times as much, so that if the same boilers which originally could just furnish steam with wood or coal were attached to a furnace burning tan the chimney would have to be much higher.

AIR SUPPLY.

The least air used in any of the experiments when the combustion was complete was 10 pounds per pound of dry tan, or an excess over that actually necessary of 40 per cent., about the same as is found in a coal fire.

RESULTS OBTAINED.

That when these proportions are observed the combustion was found to be perfect in the Hoyt and Crockett oven, and would undoubtedly be perfect in the Thompson if the ash pit was cleared out and the area of outlet at the back of the furnace enlarged, and that while it may be the results of the patented furnace may become with skillful management and design *equal* to the unpatented, it appears that the unpatented furnace is practically perfect, and that therefore the patented attachments have no value at all.

THERMAL EQUIVALENT OF BARK.

The numbers assumed for the thermal equivalent of bark dried at 110° C. (unleached 8½, leached 9.00 pounds of water from 212°) are probably *comparatively correct.* Nevertheless, from the fact that in the two experiments at Wilcox the heat accounted for overruns these numbers, while at Wells' (1) and (2) it reaches it, there being no margin for incomplete combustion and radiation. I am of the opinion that the thermal equivalent of hemlock bark, probably from the extra amount of resin it contains, is more than these numbers, and is at least 9 pounds of water from 212° for the unleached bark dried at 110° C. and 9½ pounds for the leached bark.

The details of the drying of the tan, and also of a very elegant experiment for drawing and analyzing the chimney gas, devised and executed by my associate, Mr. J. K. Rees, were under his especial charge, and are properly the subject of an independent report.

Very respectfully,

THERON SKEEL.

APPENDIX.

Dimensions of Ovens and Boilers.

Location of Furnace	Great Bend, N. Y.	Webb's Mills N. Y.	Br'kneyville, N. Y.	Binghamton N. Y.	Wilcox, Pa.
Name of owner	Stevens.	Wells.	......	Weed.	Schultz.
Kind of oven	Crockett.	Crockett.	Thompson.	Thompson.	Hoyt.
Kind of boiler	Flue.	Flue.	Flue.	Tubular.	Flue.
Material of chimney	Iron.	Iron.	Iron.	Brick.	Brick.
Hight of chimney above grate, ft.	74	90	80	94	109
Grate surface, sq. ft.	60	78	162	243	192
Heating surface, sq. ft.	400	830	480	2,000	950
Cross area of flues or tubes, sq, ft.	2.50	5.54	3.14	8.33	6.4
Cross area of stack, sq. ft.	5.40	5.24	4.90	9	16 for 4 ovens
PROPORTIONS.					
Grate surface to heating surface	1 : 6.66	1 : 16.4	1 : 2.9	1 : 8.2	1 : 4.75
Grate surface to cross area tubes	24 : 1	14 : 1	51 : 1	29 : 1	30 : 1
Grate surface to cross area chimney	11 : 1	33 : 1	33 : 1	27 : 1	24 : 1

Statement of Means and Totals of Observations and Measurements.

Date of Experiment	Sept.	Sept.	Sept.	Sept.	Oct.	Oct.	Oct.	Nov.	Nov.
Name of owner of furnace	Stevens	Wells	Brack	Weed	Wells	Wilcox	Wilcox	Wells	Wells
Number of experiment	I.	I.	I.	I.	II.	I.	II.	III.	IV.
Duration of experiment, hours	12	12	12	12	12	24	24	12	12
Pressure of steam	60	55	73	67	29	75	75	65	30
TOTALS.									
Pounds of wet tan consumed	20,129	26,814	10,392	36,138	17,282	62,123	46,612	17,027	16,512
Pounds of ashes and refuse	204	265	......	......	......	......	......	......	......
Pounds of combustible	19,925	26,549	......	......	......	......	......	......	......
Pounds of feed water	16,760	27,920	14,430	70,035	18,560	98,018	81,250	24,570	22,365
TEMPERATURE.									
Furnace (maximum)	......	1,8[illegible]8°	......	2,000°	1,900°	2,000°	2,000°	1,900°	1,900°
Furnace (average)	......	......	1,900°	1,900°	......	2,000°	2,000°	......	......
Atmosphere	50°	70°	50°	65°	50°	70°	70°	30°	30°
Chimney	700°	580°	580°	510°	420°	700°	580°	490°	455°
Feed water	140°	59°	140°	195°	140°	41°	41°	140°	140°
Ash pit	480° 545°	700°	1,500°	1,500°	......	70°	480°	700°	......
Per cent. water in wet spent tan dried at 110° C.	61.5	63.4	59.0	55.1	62.8	61.2	61.2	61.5	62.3
Per cent. ashes and refuse	2.6	2.7	......	......	......	......	......	......	......
Per cent. water in gr'n ground bark dried at 110° C.	......	......	......	......	......	......	fine 18.1	crs. 17.0	......
WEIGHTS.									
Weight cord unleached bark	......	2,582	2,353	......	......	fine 2418	crs 2284	......	......
Weight cord leached wet tan	4,442	4,294	4,112	4,270	4,225	4,076	4,076	4,275	4,280
Weight cord leached dry tan	1,710.2	1,571.6	1,686	1,917.2	1,571.7	1,581.9	1,581.9	1,645.9	1,606.0
Weight cord packed wet tan	......	......	5,237	......	5,490	......	......	......	......

Mean of Observations on Apparatus for Detecting Water Entrained in Steam.

Experiment	Wells. (1.)	Brackney.	Weed.	Wells. (2.)	Wilcox.
Number of hours apparatus was in use	5	5	6	6	11
Mean temperature of injection	49.4	57.25	56.29	48.8	40.9
Mean temperature of discharge	99.1	98 5	104.61	73.73	83.44
Mean temperature of worm	111	102	118.3	74.9	92.80
Pounds water from worm per hour	99	91.7	94.4	49.8	81.18
Head of water in tank, inches	19.70	21.75	18.3	19.625	20.125
Pressure of steam above atmosphere	55	53	71	26.	69

Calculated Results From Experiments.

Name of furnace	Stevens.	Wells.	Brack.	Weed.	Wells.	Wilcox.	Wilcox.	Wells.	Wells.
Number of experiment	I.	I.	I.	I.	II.	I.	II.	III.	IV.
Percentage of water in tan	61.5	63.4	59.0	55.1	62.8	61.2	61.2	61.5	62.3
Percentage of solid matter in tan	38.5	36.6	41.0	44.9	37.2	38.8	38.8	38.5	37.7
Weight of cord of wet tan	4,442	4,294	4,112	4,270	4,225	4,076	4,076	4,275	4,260
Weight of dry portion of do	1,710	1,571.6	1,648.9	1,917.2	1,571.7	1,581.9	1,581.9	1,645.9	1,606.0
Weight of insoluble portion do	1,626	1,478	1,536	1,509	1,424	1,484	1,484		
Weight of portion lost by boiling	84	93.6	112.9	408.2	147.7	97.9	97.9		
HOURLY QUANTITIES.									
Pounds of wet tan burned	1,675	2,235	865	3,011	1,440	2,588	1,942	1,419	1,376
Pounds of dry tan burned	645	818	355	1,352	536	1,004	754	546	519
Pounds of wet tan per sq. ft. grate	27.9	28.7	5.4	12.4	18.4	13.4	10.1	18.2	18.0
Pounds of dry tan per sq. ft. grate	10.7	10.5	2.2	5.6	6.8	5.2	3.9	6.1	6.0
Pounds of wet tan per sq. ft. heating surface	4.19	2.69	1.81	1.50	1.72	2.72	2.04	1.71	1.66
Pounds of dry tan per sq. ft. heating surface	1.61	0.98	0.73	0.67	0.64	1.05	0.79	0.66	0.63
Pounds of wet tan per sq. ft. cross section of flues	67?	404	270	360	262	404	303	258	550
Pounds of dry tan per sq. ft. cross section of flues	258	147	110	160	98	157	119	99	207
Pounds water evaporated	1,397	2,327	1,203	5,836	1,547	4,084	3,385	2,048	1,864
Pounds water entrained with steam	......	33	95	188	55	*s.h. 24°	*s.h. 24°	....	......
ECONOMIC PERFORMANCE—POUNDS OF WATER EVAPORATED.									
Per pound wet tan from feed	0.833	1.041	1.390	1.938	1.074	1.523	1.750	1.443	1.355
Per pound dry tan from feed	2.161	2.809	3.329	4.319	2.887	3 925	4.510	3.747	3.594
Per pound wet tan from 212°	0 893	1.117	1.680	1.988	1.192	1.790	2.058	1.605	1.480
Per pound dry tan from 212°	2.319	3.025	4.098	4,529	3.204	4.613	5.304	4.168	3.954
Pounds water evaporated from 212° per sq. ft. heating surface	3.79	3.36	3.00	2.92	2.25	5.19	4.26	2.73	2.50
Per cent. of carbonic acid in dry gas by volume	......	7.5	8.9	11.1	.....	......	......	8.2	....
Per cent. of oxygen in dry gas by volume	......	14.9	13.3	10.6	....	......	.....	14.7	.....
Pounds air per pound dry tan calculated from velocity of smoke	......	22	18	10½	.....	10	10	16	.....
DISTRIBUTION OF HEAT OF ONE POUND OF DRY TAN IN POUNDS OF WATER EVAPORATED FROM 212°.									
Gas passing up chimney heated from temperature of atmosphere to temperature of chimney	......	2.94	2.40	1.33		1.73	1.40	1.96	.....
Water in tan evaporated from 68° and superheated to temperature of chimney	......	2.28	1.91	1.61	.....	2.19	2.09	2.07	.....
Water evaporated in boiler from 212°	......	3.03	4.01	4.53	.....	4.61	5.30	4.17	.....
Water of combustion, estimated	......	0.45	0.45	0.50	.....	0.45	0.45	0.45	.....
Sum of above four quantities	......	8.70	8.77	7.97	.....	8.98	9.24	8.65	.....
Thermal equivalent of tan (cal)	......	9.0)	8.9	8.60	.....	9.00	9.00	8.90	.....
Difference, being loss from all causes, incomplete combustion, radiation, etc	......	0.30	0.13	—0.63	.....	0.02	0 24	—.25	.....
Available heat per pound of dry tan	......	6.72	6.99	6.99	.....	6.81	6.91	6.83	.....
Absolute efficiency of furnace and boiler	......	45.0	57.3	64.7	.....	67 8	76.7	61.1	.....

*Super-heated, 24°.

REMARKS ON EXPERIMENTS.

(From Note Book.)

Made an experiment on Crockett furnace Sept 21 and 22. Used bark that had been pitched out from leaches 24 hours before, and had been draining on ground during that time. Tan was measured in box and each box weighed. Furnaces were fired by M. A. Stevens in the usual manner, *i. e.*, furnace was fired every 10 minutes nearly,, tan being thrown in through both doors and whole surface of grate covered 6 inches or 8 inches thick. Considerable white smoke issued from the stack all the time, but most for a few minutes after firing. Temperature of crown of furnace near fire bridge (perhaps over half of furnace) was, when the door was opened, a dull red; soon became black after firing. A part of unburned tan dropped through holes in grate into ash pit at a bright red heat and slowly consumed there without blaze. The temperature of ash pit was found by thermometer held close up under grate bars, near back end, 480°–545°. The grates were of iron and had been in use 8 years; were badly warped, and cracks were covered over with fragments of iron; it was through these cracks that tan fell into ash pit. There must have been considerable heat lost by radiation from brick work and iron uptake. Chimney was part brick and part sheet iron. Sparks could be seen in large quantities through hole cut for thermometer, and *once* were seen from top of stack. A great many sparks seemed to leave the furnace over bridge wall, and on opening back connection doors cinders were found which commenced to consume as soon as air was admitted. * * * Feed water was heated by passing through pipe surrounded by steam; was found to be *nearly tight.* Temperature of feed 150°, well water 54°. Tally of barrels of water and boxes of tan were kept by two persons independently. Water was started at three cocks and brought to this point at end. Sample of tan was taken from various parts of heap at 8 p. m. Zinc was not melted in chimney or ash pit. Copper was not melted in furnace. Silver was melted in furnace. There were many crevasses in brick work around boiler which emitted smoke during some portion of time, and which in back connection were surrounded by deposit of "coal tar." Dimensions of box in which tan was measured: 3 feet 6⅓ inches×2 feet 3 inches×2 feet 0½ inch; capacity, 16.2 cubic feet.

* * * * * * * * *

Made experiment on Crockett furnace at Webb's Mills, recently erected, and owned by Mr. Wells. Cone grates in good condition. * * * Water measured in barrels, tan measured in boxes. Tan had been draining in leach 24 hours before and was pitched during experiment; was wheeled 100 feet in barrows, and then dumped 10 feet to fire room floor. * * * During first part of experiment damper one-half open and furnace cool; ash pit door 8 inches open. During last half fires thicker, *i. e.*, 6 inches over crest and 15 inches over hollow of bars; damper open; ash pit doors 2½ inches open; furnace much hotter; steam variable, from 45 to 65 pounds. Stack of sheet iron. Feed water passed through heater said and believed to be tight. Pumps worked all time. Water supplied to barrels by centrifugal pump. Sides of furnace and boiler enclosed by walls of house. Tan apparently wetter than at Great Bend and coarser. Boilers apparently good. Gases drawn every hour. * * * Engine running all night, blowing off occasionally. Water carried during experiment at first gauge, but commenced and ended at second gauge. Arches were noticed toward the end to be *bright* red before firing. Seams below the surface remaining hot when arches were black after firing. Weight of a barrel of water, 315 pounds. Dimensions of box for measuring tan, 3 feet 2 inches×2 feet 3¼ inches×1 foot 11 inches; capacity, 13.78 cubic feet.

* * * * * * * * *

Made experiment on furnace at Binghamton, Sept. 29, 9 a. m. till 9 p. m. Ovens were fed with tan that had drained 36 hours, and was pitched continuously during experiment. Tan weighed and measured as in other experiments. Water measured in casks. The weight of water at 56° necessary to fill cask, 358 pounds. Temperature of water when measured during experiment, 205°. Had counter on pump. Pump delivered a varying volume of water, depending upon the depth of water over the suction pipe. Priming determined in the usual way. Air clear, wind south and fresh. No smoke from the stack at any time. Never melted copper in combustion chamber. Silver melted easily; silver melted in ash pit when held close to surface of embers. Ash pits were not cleaned during the experiment except No. 2 oven. Back connection doors showed dull red in the dark. Dimensions of box for measuring tan, 2 feet 8 inches×2 feet 10 inches×2 feet 4¼ inches; capacity, 17.8 cubic feet.

Dampers wide open all the time. No smoke from chimney. Time of smoke reaching chimney top from combustion chamber, No. 1 and 2 oven, 10 seconds.

* * * * * * * * *

Made experiment on Thompson furnace, Sept. 25-26, from 8 p. m. till 8 a. m. Tan used measured in a box and weighed on scales. About one-third of the tan had been over the oven some hours (at least twelve) before experiment commenced; balance shoveled from the leach about eight hours before. These two kinds of tan were mixed by shoveling from one end of the fire room to the other before experiment commenced. Whenever weight of box of tan in log is marked (*) signifies that a boxfull had been taken from top of oven; others were taken from heap not on top of oven. Ovens were kept buried about three feet deep in wet tan during whole experiment. Ash pits were not cleaned out, as owners said they could not keep up steam without cinders in them. Boilers connected to brick pier supporting chimney through sheet iron front connection. Wet tan laying against this connection took fire and burned with flame during whole experiment. Made sure that the blow-off valve was not leaking by filling it with tan. Silver was melted in one flue leading from one pair of ovens but not from other. Fires on being examined *seemed* to be equally hot. Furnace was fired every 2½ hours, nearly, through holes furthest from the door, and every 1½ hours through holes nearest to door. All holes in all furnaces in same position were fired at once, *i. e.*, all holes nearest to boiler every $2\frac{1}{2}$ hours, and all holes furthest from boiler every $1\frac{1}{2}$ hours. Furnace doors had no holes in them and were kept tightly closed and luted with fire clay. Ash pit doors were of sheet iron, about 2 feet square, and were kept so as to expose an opening for the passage of air of about 40 square inches to each oven. Glass water gauge in chimney showed one-half inch; did not fall when large door to back uptake, 2 square feet, was opened. Temperature of stack, 580°. Seems probable that almost all combustion took place on surface of cone. Amount consumed in ash pit small. Temperature of ash pit 600°. Water was measured in barrels and passed into tank by three men with pails. Tally was kept by two men independently. Chimney gases were drawn every hour till 4 a. m. Experiment on velocity of air in stack was made with coal smoke. Time required to reach chimney top from combustion chamber was: 10, 11, 10, 11, 12, $11\frac{1}{2}$, 9, 9, 9, 9, 9 seconds; mean, 10.04 seconds. Weight of a barrel of water, 390 pounds. Flues were swept before experiment commenced. Boilers were fed by injector; no heater; drip water from injector returned to tank. Doors around the furnace and cracks were all luted with blue clay and ashes before experiment. Holes above the furnace were kept covered with tan, and covers not used. Weather rainy during the night and damp in morning. Air, 50°; boiler room, 70°. Dimensions of box for measuring tan: 2 feet $1\frac{1}{4}$ inches$\times$2 feet $8\frac{1}{8}$ inches$\times$2 feet $7\frac{1}{4}$ inches; capacity, 14.69.

THERON SKEEL.

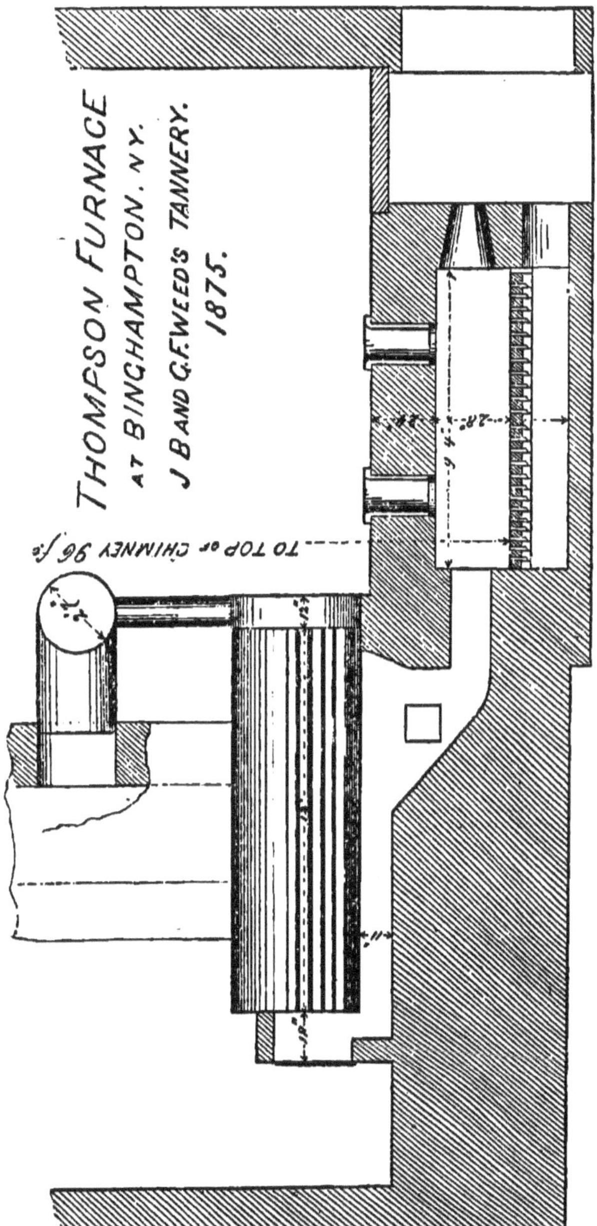
THOMPSON FURNACE
AT BINGHAMPTON, N.Y.
J B AND G.F. WEED'S TANNERY.
1875.
TO TOP OF CHIMNEY 96 ft.

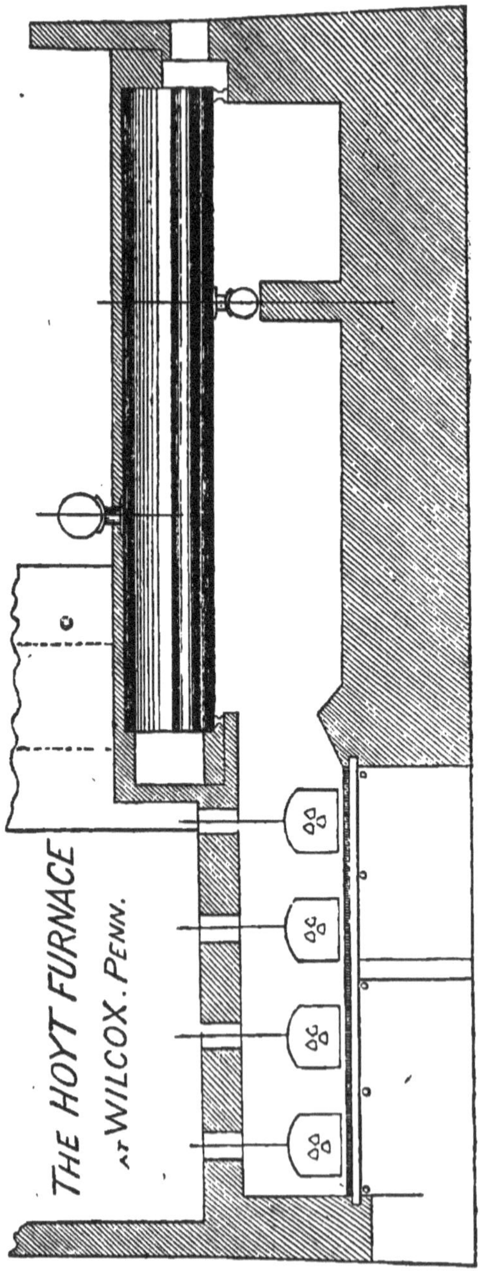

THE HOYT FURNACE AT WILCOX, PENN.

CROCKETT FURNACE AT WELLS TANNERY.
1875.

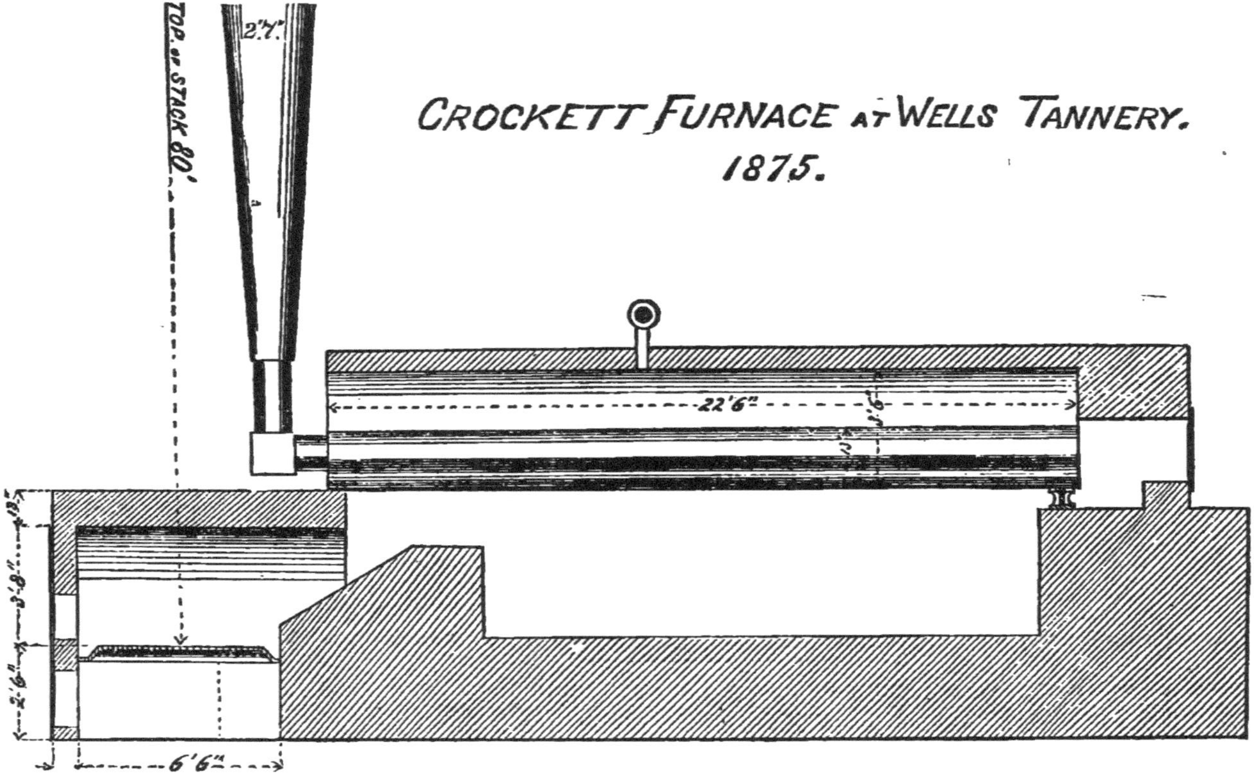

FIG. 1.

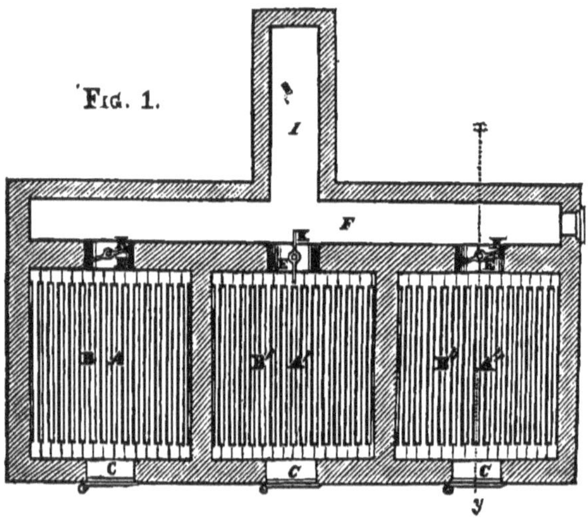

FIG. 2.

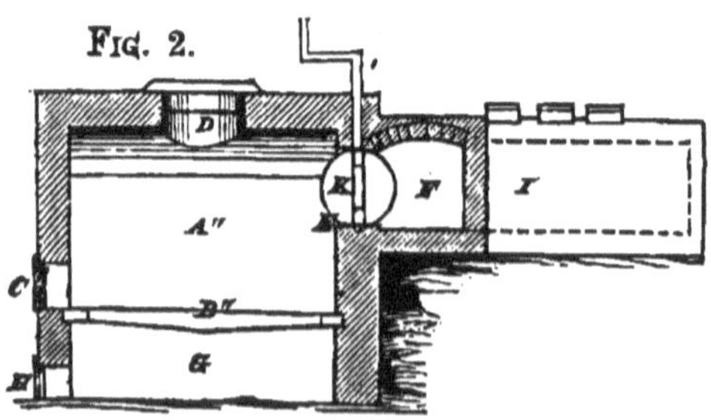

THOMPSON FURNACES.

The Moses Thompson Patent for Wet Tan Burning Furnaces.

(Extracts from Patent, showing Claim and manner of working Furnaces.)

SUBMITTED BY J. S. SCHULTZ, J. B. HOYT AND OTHERS.

Fig. 1 is a horizontal section of a furnace constructed according to my invention.

Fig. 2 is a vertical section of the same in the line *x y* of Fig. 1.

* * * * * * * *

The main object of my invention is to effect the more economical use for fuel of tan bark, bagasse, or other trashy matters in a wet state, or very green or wet wood.

It is also applicable to the burning of fuel of that or other descriptions in a dry state; but when the fuel is dry it is much less useful, and I do not claim its application thereto.

The nature of my invention consists in the employment, in the manner hereinafter described, of a series of fire chambers arranged side by side or in any convenient way, to admit of the whole series communicating with the same heating flue, which said fire chambers are furnished with dampers by which their respective communications with the flue and the ash pit may be closed or opened at pleasure.

This arrangement is for the purpose of enabling the process of heating the fuel to an intense degree in a nearly air-tight chamber, and then admitting a free supply of air to promote its rapid combustion, to be conducted without interruption to the operations which the heat generated is intended to effect.

To enable others skilled in the art to make and use my invention, I will proceed to describe its construction and operation.

The furnace shown in the drawings has three fire chambers, A, A ′, A ′′. Three is the number shown, as I consider that number in most cases to be the best adapted for practical operation. The fire chambers are of square but may be of other form with grate bottoms, B, B ′, B ′′, and arched tops. They are separated by walls of non-conducting material, and lined throughout with fire brick. Each is provided with a door, C, in front for the purpose of lighting and tending the fire, with an opening, D, at the top, for the purpose of supplying the fuel when it consists of tan bark, saw dust, or other material of a similar nature, and with an opening, E, at the back, which leads to the flue, F, the opening, E, being provided with a damper, K. Each fire chamber has a separate ash pit, G, below it, which is furnished with a door, H, to regulate the admission of air.

The flue, F, extends across the back of all three fire chambers, and the chimney may be at one end or may be placed in the rear, with a flue, I, leading to it from the flue, F. If the furnace is used for generating steam, the best place for the boiler will be in the flue I, which will be made of proper size to receive and nearly surround it. If used for other purposes, any arrangement may be made that may be considered best, but the thing to be heated ought to be so high as not to require the products of combustion to descend on their passage to it.

The mode of conducting the operations of the furnace is as follows:

Fires being lighted in all the fire chambers, two of the three have the doors, H, H, of their ash pits closed, and the dampers, K, K, so nearly closed as only to allow a sufficient escape of the gases generated by the slow combustion which then goes on to prevent explosion. The other fire chamber in the meantime has the damper K open, and the door of the ash pit opened far enough to admit any quantity of air that may be requisite to promote such a degree of combustion in the chamber as may be necessary to generate the amount of heat required. The air should be drawn in (except when excluded) by natural draft, and if a high stack be used, there should be a damper in it to check the draft.

When the fluel in the open chamber is reduced to a desirable degree that chamber is closed and recharged, and another opened and supplied with air until the fuel within is reduced, when it is closed, recharged, and another opened, each in its turn being opened and freely supplied with air to generate and supply the requisite amount of heat, while the others are closed and successively supplied with fresh wet fuel to heat and decompose the same to such a degree as is desirable before allowing rapid combustion and escape of the heat to take place.

Each fire chamber should be supplied successively with fuel at proper intervals by any convenient means, either through the hole, D, or through the door, C, in front, just before closing the fire chamber.

The principal advantage of a furnace and process of this description consists in heating and decomposing the fuel without any further loss of heat than is absorbed by the poor conducting material of which the furnace is constructed, to such a degree as will, when a proper supply of air is admitted, cause the most perfect combustion of the gases and smoke to be effected. This could not be effected in a single fire chamber without interruption to the proper supply of heat, but when two or more fire chambers are employed no interruption takes place, as one furnace is always in full operation. Another advantage consists in always holding a certain quantity of heat in reserve in the closed chambers, which may be immediately brought into action by opening one or more of the chambers.

A similar but inferior result might be produced by having several separate grates and ash pits to the same fire chamber, each grate charged successively, and its ash pit for a time closed, immediately after fresh charging, to exclude the air.

I have described this in my caveat upon which this application is based, but do not use it because of its inferiority in practice, although it involves my principle.

After ample experiments, I have discovered that any results which can be produced by the use of dry fuel are entirely inferior to mine in proportion to the quantity used, and that results like mine can only be attained by the use of wet fuel, similar to what I have herein mentioned, fed into an intensely heated chamber.

Under such circumstances, the water in the fuel, in the presence of the carbonaceous substances in the furnace, will be decomposed, giving its oxygen to the carbonaceous matter, dispensing with a draft and its cooling and wasteful influences, and rendering the combustion so perfect that no smoke is visible.

I do not claim the within-described arrangement of a series of fire chambers to communicate with one common flue, irrespective of the purpose for which, and the manner in which, I employ the said arrangement.

But what I do claim as my invention, and desire to secure by letters patent, is :

The combustion, for the purposes of a high degree of heat, of bagasse, refuse tan, saw dust, and other refuse substances, or very wet and green wood, by the employment of a series of fire chambers, arranged in any manner substantially as described, to communicate with one flue, when any number of the said chambers are nearly closed to the flue and to the admission of air, when first charged as described, while the remaining chamber or chambers is in free communication with the flue, and has a free supply of air admitted, and each chamber in its turn is nearly closed, and then opened, and has air admitted, whereby the heat required is furnished by the combustion of the fuel in one or more chambers, while the fuel in the other chamber or chambers is being heated and decomposed to a desirable degree, as herein set forth, no artificial blast being used.

NOTE.—The drawings of the real Thompson furnace, as seen on page 58, and as described on the three following pages, will sufficiently indicate the original idea of the patentee. But the claim has been construed by the Court to cover the structure on page 55. For the purposes of this examination the legally construed furnace has been assumed to be the true Thompson furnace. There are not over three or four real Thompson furnaces—as indicated by the original drawings and description—in existence, and they give very inferior results. J. S. SCHULTZ.

www.ingramcontent.com/pod-product-compliance
Lightning Source LLC
Chambersburg PA
CBHW032045220426
43664CB00008B/862
* 9 7 8 3 3 3 7 1 0 6 4 7 8 *